BIOETHICS IN ACTION

Speaking from and to the growing movement among academics to become involved with "socially engaged" work, this volume presents first-person case studies of attempts to fix serious ethical problems in medical practice and research. It highlights the critical difference between the pundit approach to bioethics and the interventional approach – the talkers and the doers – and points to how abused and damaged the doers often end up. Chapters cover a diverse set of topics, including the troubling influence of for-profit businesses on public health policy, the politics of exposing histories of unjust medical research, the challenges of patient rights' work in sexuality and reproduction, collaborations between NGOs and academics, methods for changing entrenched yet harmful medical practices, engaging public policy through educating governmental leaders, and whistleblowing. The trending interest in the interplay of academia and advocacy and the growing importance of "socially engaged" work by academics make this a timely and much-needed resource.

FRANÇOISE BAYLIS is an internationally renowned bioethics expert whose innovative work, at the intersection of policy and practice, has stretched the very boundaries of the field. Her ethics research focuses primarily on women's reproductive health and genetic technologies. Baylis believes bioethicists need to exercise their moral imagination and find creative ways to make the powerful care. She is a member of the Order of Canada and the Order of Nova Scotia and a fellow of the Royal Society of Canada and of the Canadian Academy of Health Sciences. In 2017, she was awarded the Canadian Bioethics Society Lifetime Achievement Award.

ALICE DREGER is an historian of science, medicine, and sexuality, a patient advocate, mainstream writer, and public intellectual. Her most recent book, *Galileo's Middle Finger*, funded by a Guggenheim Fellowship, has won praise in the *New Yorker*, the *New York Times*, *Nature*, *Science*, and beyond. An award-winning writer, Dreger has published in many major venues including *WIRED*, *Slate*, the *Guardian*, the *Wall Street Journal*, *New Statesman*, the *Atlantic*, and the *New York Times*, and her TED talk, "Is Anatomy Destiny?" has been viewed over a million times. *The Chronicle of Higher Education* has called Dreger a "star scholar."

CAMBRIDGE BIOETHICS AND LAW

This series of books was founded by Cambridge University Press with Alexander McCall Smith as its first editor in 2003. It focuses on the law's complex and troubled relationship with medicine across both the developed and the developing world. Since the early 1990s, we have seen in many countries increasing resort to the courts by dissatisfied patients and a growing use of the courts to attempt to resolve intractable ethical dilemmas. At the same time, legislatures across the world have struggled to address the questions posed by both the successes and the failures of modern medicine, while international organisations such as the WHO and UNESCO now regularly address issues of medical law.

It follows that we would expect ethical and policy questions to be integral to the analysis of the legal issues discussed in this series. The series responds to the high profile of medical law in universities, in legal and medical practice, as well as in public and political affairs. We seek to reflect the evidence that many major health-related policy debates in the United Kingdom, Europe and the international community involve a strong medical law dimension. With that in mind, we seek to address how legal analysis might have a trans-jurisdictional and international relevance. Organ retention, embryonic stem cell research, physician assisted suicide and the allocation of resources to fund health care are but a few examples among many. The emphasis of this series is thus on matters of public concern and/or practical significance. We look for books that could make a difference to the development of medical law and enhance the role of medico-legal debate in policy circles. That is not to say that we lack interest in the important theoretical dimensions of the subject, but we aim to ensure that theoretical debate is grounded in the realities of how the law does and should interact with medicine and health care.

BIOETHICS IN ACTION

Edited by

FRANÇOISE BAYLIS

Dalhousie University, Halifax, NS, Canada

ALICE DREGER

CAMBRIDGE
UNIVERSITY PRESS

CAMBRIDGE
UNIVERSITY PRESS

University Printing House, Cambridge CB2 8BS, United Kingdom

One Liberty Plaza, 20th Floor, New York, NY 10006, USA

477 Williamstown Road, Port Melbourne, VIC 3207, Australia

314–321, 3rd Floor, Plot 3, Splendor Forum, Jasola District Centre, New Delhi – 110025, India

79 Anson Road, #06–04/06, Singapore 079906

Cambridge University Press is part of the University of Cambridge.

It furthers the University's mission by disseminating knowledge in the pursuit of education, learning, and research at the highest international levels of excellence.

www.cambridge.org
Information on this title: www.cambridge.org/9781107120891
DOI: 10.1017/9781316343197

First published 2018

Printed in the United Kingdom by Clays, St Ives plc

A catalogue record for this publication is available from the British Library.

ISBN 978-1-107-12089-1 Hardback

CONTENTS

CONTRIBUTORS

FRANÇOISE BAYLIS, Novel Tech Ethics, Faculty of Medicine, Dalhousie University, Halifax, NS, Canada

JOCELYN DOWNIE, Health Law Institute, Dalhousie University, Halifax, NS, Canada

ALICE DREGER, author of *Galileo's Middle Finger: Heretics, Activists, and One Scholar's Search for Justice*

JOEL E. FRADER, Department of Pediatrics, Feinberg School of Medicine, Northwestern University, and Lurie Children's Hospital, Chicago, IL, USA

RUTH MACKLIN, Department of Epidemiology and Population Health, Albert Einstein College of Medicine, Bronx, NY, USA

JANE MAIENSCHEIN, School of Life Sciences, Arizona State University, Tempe, AZ, USA

SUSAN M. REVERBY, Women's and Gender Studies, Wellesley College, Wellesley, MA, USA

ARON SOUSA, Department of Medicine, Michigan State University, East Lansing, MI, USA

MIRIAM ZOLL, Emeritus Board Member, Our Bodies Ourselves, and author, *Cracked Open: Liberty, Fertility and the Pursuit of High-Tech Babies*

ACKNOWLEDGMENTS

Sincere and effusive thanks are owed to Timothy Krahn for his care and attention to details at the peer review stage and for his indefatigable assistance with referencing. We owe him a debt of gratitude. We are also grateful to the contributors who stuck with us as we worked for years to bring their stories together in a way we hope is maximally accessible and powerful. In particular, we thank Carl Elliott and Eric Suba for their testimonials that, ultimately, we were not able to include in this collected edition. As well, we are also very grateful to the staff of Cambridge University Press, particularly Finola O'Sullivan, for shepherding us through the challenges faced by a volume of this type.

This work was supported in part by the Canada Research Chairs program (2011–2018) on Impact Ethics. Funding was provided by the Canadian Institutes of Health Research to Françoise Baylis for an invitation-only workshop in Halifax in the fall of 2012. The purpose of the workshop was to discuss ways and means of promoting ethical policy in the fields of health, science, and biotechnology. We sought to engage key stakeholders and knowledge purveyors (bioethicists, policy analysists, and journalists) in debate and discussion about ways of promoting innovative, responsible, and accountable bioethics as a means to effect the development of sound public policy. There were many positive outcomes from this workshop, one of which is the blog Impact Ethics (which is thriving). Another positive outcome is this collected edition. Many of the contributors to this book were not participants in the original workshop, but the germ of an idea for this book project was sowed at this inaugural meeting. As well, protected research time to work on this project was made possible by generous support of the Brocher Foundation, www.brocher.ch.

1

More than Words

ALICE DREGER AND FRANÇOISE BAYLIS

This work is borne of shared frustration: frustration over seeing many specific instances of bad behavior, poor judgment, dangerous ignorance, harmful cultures, or outright injustices in contemporary medical care and medical research; frustration over medical systems punishing whistle-blowers and agents of positive change while rewarding bad actors and interest-conflicted parties; frustration over the public being distracted by sexy bioethics micro-punditry while so much harm is accruing, largely unacknowledged, at a much larger scale; frustration especially at how "bioethics" has come to be understood – within the academic realm as something that is increasingly measured by external grants and (secondarily) peer-reviewed publications, and within the public realm as sporty commentary on oversimplified medical controversies purposefully dramatized to increase media outlets' ad revenues.

The bioethics we have known personally, in our own work and our collaborations, is largely different from this. A few years ago we labeled our kind of bioethics "Impact Ethics." In our conception, Impact Ethics is fundamentally oriented toward public service, and, as such, it often challenges the "powers that be" in a direct fashion. Consider, for example, an early, apt meditation on Impact Ethics penned by our colleague Barry Hoffmaster:

> Impact Ethics is a way of doing bioethics.
> It is not a definition of bioethics.
> It is not a theory of bioethics. It is not a scholarly analysis of bioethics.
> It is a bioethics that is faithful to its origins, not its patrons.
> It is a bioethics that is embedded, not abstracted.
> It is a bioethics that is critical, not accepting.
> It is a bioethics that is creative, not conventional.
> It is a bioethics that is risky, not secure.
> It is a bioethics of commitment, not resignation.
> It is a bioethics of engagement, not passivity.
> It is a bioethics of devotion, not recognition.
> It is a bioethics of the marginalized, not the established.

> It is the bioethics of a calling, not a career.
> It is the bioethics of the advocate, not the consultant.
> It is the bioethics of the activist, not the apologist.
> It is the bioethics of the whistleblower, not the muckraker.
> It is a bioethics for the humble, not the aggrandizing.
> It is a bioethics for the vulnerable, not the powerful.
> Impact Ethics disappoints more than it rewards.[1]

Why does Impact Ethics "disappoint more than it rewards"? Because it is the kind of work that pushes against powerful forces and against the reward systems those forces have established to protect themselves. The work of Impact Ethics does not lack academic research and theorizing, but it is so much more than this. The outcome(s) that matter are not lines in a resume, dollars in research accounts, numbers of appearances on television or talk shows, or service activities that primarily aim to please one's university administration. That is, the outcomes that matter are neither career advancement nor popularity. The outcomes that matter are witnessing and ending injustices, improving patients' lives, defending and protecting research participants' rights, and contesting entrenched (and too often accepted) structural conflicts of interests. This is often, as this volume shows, deeply uncomfortable and personally costly work; Impact Ethics is not easy work.

Why would anyone take this on? As the contributions to this volume show, there are a variety answers to this question. Some of us have attempted ethical interventions in medicine and medical research because we personally experienced an injustice that we realized was part of a larger web of injustice also harming others. Some of us have attempted ethical interventions because we discovered something in our research that we felt should not go unaddressed. Some of us are naturally inclined toward social and political engagement. And some of us have personality types that make it difficult for us to let go when we see harmful systems that we feel require public exposure, accountability, and change.

What all of the stories in this volume have in common are attempts to approach ethics not as something to just *talk about*, but as something to *do*. Paradoxically, this is what people outside the academic field known as bioethics often assume is what the field of bioethics *normally* involves; after all, bioethics *is* applied ethics. Indeed, many physicians, nurses,

[1] B. Hoffmaster, "What is impact ethics?," *Impact Ethics*, November 18, 2013, available at: https://impactethics.ca/2013/11/18/what-is-impact-ethics/. Accessed on: February 21, 2018.

other healthcare professionals, patients, research participants, and their families assume that what academic and clinical bioethicists do through words and deeds *is* advocacy on behalf of patients and medical research participants. Indeed, it is not uncommon for academic and clinical bioethicists to be approached by individuals hoping they will be of help with actual medical ethics problems, policy gaps, or injustices actively occurring within the modern medical-research industrial complex.

But in practice, many bioethicists do not actively work to defend or protect specific vulnerable individuals or groups. Some theorize and write about justice but are of the view that engaging in translational or direct interventional work is the role and responsibility of others, such as lawyers, investigative reporters, professionalized activists, and administrators charged with ethics oversight. Other bioethicists advance ideas on specific topics and then, in the wake of controversy concerning their theorizing, have the audacity to suggest that they were merely exploring ideas and did not mean for their ideas to influence the "real" world of patients, families, and health-care providers.

Yet bioethics has also naturally drawn to its venues people like the authors in this volume, people we might call *doers*. These are people who wish to try to *enact*, and not just *discuss*, justice – often with a specific concentration on social justice concerns – in medicine, medical research, and public health. These include people who are primarily academics in bioethics and allied fields like health law, academic medicine, and history of medicine and science, but also those who are public sector advocacy lawyers, nonprofit activists, and investigative journalists. These are people who have, by their orientation, grown increasingly intolerant of people called bioethicists who advocate *not* primarily on behalf of patients or research participants but on behalf of those who fund the bioethicists – in some cases, drug, device, or vaccine makers, in other cases, research groups empowered by and empowering the medical-industrial research complex.

As a consequence of all of this, an active schism – or at least an uncomfortable implied debate – appears to be forming within academic bioethics over the question of what the roles and responsibilities should be for those who are identified as bioethicists. A sign of this rift is a fundamental divergence in views about how and what bioethics should contribute to public discourse and debate, practice, and policymaking. The different perspectives reflect a fundamental contrast between what bioethics has predominantly become in academia and what bioethics could be in the world.

This emerging tension in bioethics is more broadly reflected in our culture at large, particularly with regard to the roles of academics, healthcare professionals, journalists, politicians, and other thought leaders. The following question is of pivotal importance. Working in fields where recognizing injustices is expected as part of the knowledge required for being a professional, is it enough to aim at being a competent knowledge-producer or is there an attendant duty to act (when possible) on this knowledge, in a committed and sustained fashion?

The contributors to this book share in the belief that producing knowledge is not enough. Their writings begin and end with the assumption that, amid the plurality and often conflicting needs of the various parties involved with modern medicine and medical research, the needs of actual and future patients and medical research participants must be made primary, and their rights must be respected, preserved, and even enhanced. The contributors to this book also share the view that too many people are going unprotected, misunderstood, used, and even abused in the modern medical-research industrial complex, and that it is imperative to work, in small and large – local and global – fashions to improve the clinical and research systems that are centered on the bodies of sometimes vulnerable people.

The institutionalization of bioethics in its providing of structures for keeping the needs of patients and medical research participants visible, along with institutional mechanisms to ensure respect for their rights, may contribute to this moral advance. But the institutionalization of bioethics has in practice also had the effect of diluting, inhibiting, and constraining respect for this fundamental moral requirement. Systemic, institutionally sanctioned inattention and even corruption is not uncommon in medicine and medical research. The contributors to this book insist that academic bioethics itself be held accountable for what it is and isn't doing. Moreover, the field needs to resist the all too common tolerance, or even support, for careers made merely through analysis and commentary – a kind of intellectual (and sometimes public) punditry that can feel like secondary abuse of patients and research participants who have already been harmed or are at risk of future harm.

***

In providing this topic-diverse collection of first-person narratives of bioethics in action, our goals as editors are multiple: to provide models, inspiration, warnings, and hard-won insights for those who might do the kinds of work the authors do; to issue a moral calling to those with enough privilege to act so as to protect and defend patients and medical research participants; and to better educate the public on the differences and relationships between what might be called punditry bioethics and bioethics-based advocacy.

The stories that comprise this volume are an intentional sampling of the kinds of work people are doing in specific instances and specific institutions as they try to make known and also fix medical ethics failures. By no means is this volume meant to be understood as the only example available; there is much good work being done in this area, and in some ways it was difficult to decide whom to invite to contribute. The contributors to this volume are purposefully drawn from many walks of life, including academic bioethics, history, and philosophy, but also public sector health advocacy, medicine, and politics. Some have had their stories covered in major media outlets, but this is the first time a group like this has come together to present what we see as a gripping collection of personal stories – some of failure, some of success, many of hardship. The stories presented in this collection include not only riveting accounts of medicine or medical research gone awry, but also honest reflections regarding which techniques worked or didn't work, how action might have been smarter or more effective, and where individuals (including the authors) and systems failed. As such, these stories are both inspirational and cautionary, not always in equal measure.

In Chapter 2, Françoise Baylis and Jocelyn Downie, scholars in philosophy, bioethics, and health law/policy, detail how they attempted to undo the appointment of a pharmaceutical representative to the Governing Council of the Canadian Institutes of Health Research (CIHR), the Canadian national funding agency for health research. Along with international leaders in patient advocacy, Baylis and Downie recognized the appointment as involving an untenable structural conflict of interest given the Governing Council's mandate to "shape the strategic directions, goals, and policies of federally-funded health research in Canada." The authors explain how they attempted to mobilize elected officials, fellow bioethicists, leaders within CIHR, and the public in an effort to stop having corporations with seriously checkered ethics histories from literally having a seat at the table of national health research policymaking.

In Chapter 3, Susan M. Reverby, a historian of medicine, explores how she has tried to produce history that can meaningfully inform our contemporary health-care systems. Reverby is a leading historian of the infamous Tuskegee syphilis study and was the researcher who discovered the US government's studies involving the intentional infection of prisoners with gonorrhea and syphilis in Guatemala. Reverby explains how she has tried to make known her findings of unethical medical research while avoiding unnecessary harm to public health institutions – not an easy scene when dealing with the modern media machine, which loves to simplify and sensationalize stories involving unethical behaviors. She also explores the politics of public apologies and the problem of how history of the dead is often reframed for the glory – rather than for the edification – of those living. She concludes that "doing the best history we can matters, but this will never be enough," and suggests what more must be done.

In Chapter 4, Alice Dreger, a historian and philosopher of anatomy, explains how she came to be a leader in the intersex patient rights movement, including as one of the heads of the Intersex Society of North America. She recounts what techniques she and her collaborators used to try to bring about a more rational and just system of care for children born with body types that do not fit sex norms, and she carefully points out where they succeeded and failed.

In Chapter 5, Ruth Macklin, a philosopher and bioethicist, details how she has actively collaborated with the nonprofit organization Public Citizen's Health Research Group in the United States to push for the rights of participants in clinical trials conducted nationally and internationally. Macklin shows how difficult it can be to bring about the alignment of research practices with what we generally think of as long-agreed-upon core principles.

In Chapter 6, Miriam Zoll explains how in her experience the reproductive medicine system is "a profit-driven medical marketing trap, a cultural trap, a trap of misinformation; a trap that intentionally preyed upon the naiveté and vulnerability of anxious couples." From this revelation, she explains how she moved on to become a vocal and highly effective critic of the "global repro tech industry." We include Zoll's story not only to show how iatrogenic trauma can lead to patients fighting back, but also because we believe it is important to remember that much of medical advocacy begins with patients who become politically conscious enough to wake others who may then assist. The chapters by Dreger, Macklin, and Zoll each cover decades of work and showcase how the work of challenging the medical industrial complex is not getting easier, even though it does seem to be becoming more urgent.

Chapters 7–9 provide first-person case histories of academics who have tried to help develop more educated, more compassionate, and more just systems by working within hospitals, medical schools, and governmental systems, including regulatory agencies, the judiciary, and legislative bodies. In each of these instances, the authors have identified knowledge and/or practice failures that might be fixed through direct, positive engagement. All also recognize that trying to effect this kind of systematic change requires novel collaborations and sustained efforts that ultimately require deep support from administrators who control reward systems, resources, and missions.

In Chapter 7, physician and ethicist Joel Frader tells the story of how institutionally challenging it has been to establish pediatric palliative care. His is a classic story of how good intentions can get in the way of optimal patient care, in this case in hospital settings. Frader demonstrates the kind of sustained effort required to enact meaningful change in recalcitrant systems. In Chapter 8, historian and philosopher of science Jane Maienschein describes how a highly supportive administrative system at Arizona State University in the United States has empowered her to do socially engaged work, educating judges and legislators about embryology (relevant to abortion debates), cloning (relevant to potential legislative bans on research and technology), and the like. Maienschein describes how she has set up an educational system around this outreach work in order to expand and sustain this kind of public service. Finally, in Chapter 9, physician and medical educator Aron Sousa tells the story of how he worked as a medical school dean at Michigan State University in the United States to support his faculty who were engaged with the Flint Water Crisis – a massive and protracted public health emergency caused by bad politics around drinking water. Sousa's account demonstrates how difficult it can be to manage research, care of patients, and public health when working within systems thick with politics, persistent inequities, and bureaucracy.

We are sure readers will see, as we do, that there are many questions raised by the personal narratives included in this collection. What is the difference between what is called "advocacy" and what is called "activism," besides that the former is often used as a term of admiration in medicine and public health and the latter, one of derision? What is the right way to understand academic freedom when one academic uses hers to call out another academic researcher for allegedly objectionable behavior? How do we ensure that scholarship is not weakened by ideology and missions of justice? How do we recognize that the personal is political while also understanding the limits of first-person accounts? How do we move people past simplistic stories of good and evil while also recognizing that sometimes we are dealing with cases of injustice so outrageous as to constitute a kind of evil? How do we remain morally sensitive without constant frustration and exhaustion?

We hope that this volume will spur conversations on these and many other questions with which many of us are now struggling. But even more than spurring conversation, we hope this volume conveys the sense shared by all of the contributors that actions speak louder than words.

2

"Where There's Smoke, There's Pfizer"

FRANÇOISE BAYLIS AND JOCELYN DOWNIE[1]

In late October of 2009, we stumbled across a brief online news report by Anne Silversides in the *Canadian Medical Association Journal*.[2] In this article, dated October 20, 2009, she reported that on October 5, the Federal Minister of Health had announced the appointment[3] of Bernard Prigent, a Pfizer executive, to the Governing Council of the Canadian Institutes of Health Research (CIHR) – Canada's federal health research funding agency. This was a historic appointment as it was the first time that a pharmaceutical company employee had been named to the CIHR Governing Council. As a member of the Governing Council, Prigent would not only have an opportunity, but also a responsibility, among other things, to help shape the strategic directions, goals, and policies of federally funded health research in Canada.[4]

Silversides' report included objections from Dr. Sidney Wolfe, Director of the Washington, DC – based group Public Citizen. Wolfe noted that with this appointment, the pharmaceutical industry would be able to influence policymaking to its advantage.[5] Whereas Wolfe saw this as harmful to the integrity of the research funding process, Alain Beaudet, the President of CIHR, saw this as decidedly beneficial – as a means for CIHR to "synchronize its activities with the federal government's overall

[1] Title borrowed from S. Lewis, "Where there's smoke, there's Pfizer: Sparks fly over recent CIHR appointment," *Longwoods* (December 2009). Available at: http://www.longwoods.com/content/21198. Accessed on: February 21, 2018.

[2] A. Silversides, "CIHR appointment stirs controversy," *CMAJ News* (early release, October 20, 2009). Available at: www.cmaj.ca/content/early/2009/10/26/cmaj.109-3085. Accessed on: February 21, 2018. In print: A. Silversides, "Appointment of Pfizer executive to CIHR stirs controversy," *CMAJ*, 181:11 (2009), E256–E257.

[3] An order in council appointment is an appointment by the federal Cabinet.

[4] Canadian Institutes of Health Research Act S.C. 2000, c.6, ss.14–19. Available at: http://laws-lois.justice.gc.ca/eng/acts/C-18.1/FullText.html. Accessed on: February 21, 2018.

[5] A. Silversides, "CIHR appointment." A. Silversides, "Appointment of Pfizer executive," p. E256.

science and technology strategy, *Mobilizing Science and Technology to Canada's Advantage.*"[6] He went even further and actually said, "We [CIHR and the federal government] also want to ensure closer ties with industry, that they sit at the same table at the precompetitive level. We want to help industry to succeed and encourage them to invest."[7] Whereas Wolfe was concerned about industry influence in shaping the health research agenda, Beaudet welcomed such influence.

We were shocked to read of the appointment and immediately began to try to figure out how to get the appointment withdrawn. This chapter tells the story of our largely ineffective engagement with this issue.[8]

While some might think this story is old news, events surrounding the controversial appointment are relevant for several reasons, not the least of which is that they signal the need for continued vigilance about the range of corporate strategies used[9] to influence both the conduct of health research and the delivery of health care.[10] Indeed, in early 2016, a similar scenario played itself out in the United Kingdom when a Pfizer executive – Erik Nordkamp, Managing Director and Head of Global Innovative Pharmaceutical Business UK – was named a nonexecutive director of King's College Hospital NHS Trust (an independent Foundation Trust in the United Kingdom's National Health Service).[11] With this appointment, as with the Pfizer appointment in Canada, questions were raised about "potential conflict of interest issues ... and increasing

[6] A. Silversides, "CIHR appointment." A. Silversides, "Appointment of Pfizer executive," p. E256.

[7] A. Silversides, "CIHR appointment." A. Silversides, "Appointment of Pfizer executive," p. E256.

[8] It should be noted here that this is not a story about the commercialization of research in public institutions in Canada. The potential benefits and harms of such commercialization are thoroughly canvassed and discussed elsewhere (see, for example, J. Downie and M. Herder, "Reflections on the commercialization of research conducted in public institutions in Canada," *McGill Health Law Publication*, 1:1 (2007), 23–44. Rather, this is a story about bringing the pharmaceutical industry to the center and at the highest level of decision-making about health research funding and policy in Canada.

[9] See also D. Evans, "Big Pharma's crime spree," *Bloomberg Markets*, (December 2009), 73–86. Available at: https://www.yumpu.com/en/document/view/7167355/big-pharmas-crime-spree-canadian-health-coalition. Accessed on: February 21, 2018.

[10] C. Elliott, *White Coat, Black Hat* (Boston, MA: Beacon Press, 2010). D. Healy, *Pharmageddon* (Berkeley, CA: University of California Press, 2013).

[11] E. Silverman, "Anger brewing over naming of Pfizer exec to UK foundation board," *STAT News* (January 20, 2016). Available at: www.statnews.com/pharmalot/2016/01/20/pfizer-nhs-drug-costs/. Accessed on: February 21, 2018.

pharmaceutical and private sector influence."[12] In response, Jacky Davis of "Keep our NHS Public" (a local advocacy group concerned about the pharmaceutical industry's influence on medical practice and research) insisted:

> It is not appropriate for the head of Pfizer UK to be appointed as a non-exec director at one of the country's most prestigious trusts. Clearly his presence on the board is likely to influence its attitude to the relationship between the private sector and the NHS, and increase the influence of the pharmaceutical industry within the trust and the broader NHS.[13]

The Canadian Institutes of Health Research

To appreciate the reasons for our deep concern with the Pfizer appointment,[14] and to follow our story, one must first understand the objective of CIHR, and its basic organizational structure.

CIHR was created by an Act of Parliament in 2000 in an effort to modernize the funding of heath research in Canada.[15] According to the Act, the sole objective of CIHR is "to excel, according to internationally accepted standards of scientific excellence, in the creation of new knowledge and its translation into improved health for Canadians, more effective health services and products and a strengthened Canadian health care system."[16]

It is important to emphasize this fact at the outset, because on many occasions during the debate about the Pfizer appointment, Beaudet would assert that CIHR's mandate was to facilitate commercialization. But "encouraging innovation, facilitating the commercialization of health research in Canada and promoting economic development through health research in Canada" is only one of twelve means to the laudable end of improving health for Canadians. Other explicit means to this end

[12] D. McDonald, "Big pharma now helping to run King's College Hospital?," *Our NHS* (January 20, 2016). Available at: https://www.opendemocracy.net/ournhs/diarmaid-mcdonald/big-pharma-now-helping-to-run-kings-college-hospital. Accessed on: February 21, 2018.

[13] D. McDonald, "Big pharma now helping?"

[14] We refer to this as the "Pfizer appointment," not the "Prigent appointment" (as others have done), because our concerns are not with the individual but with the corporation that employs the individual.

[15] Canadian Institutes of Health Research Act S.C. 2000.

[16] Canadian Institutes of Health Research Act, s.4.

are "promoting, assisting and undertaking research that meets the highest international scientific standards of excellence and *ethics*"[17] and "fostering the *discussion of ethical issues* and the application of *ethical principles* to health research"[18] (emphasis added).

Turning now to the governance structure, CIHR is managed by a Governing Council of no more than eighteen members.[19] The responsibilities of the CIHR Governing Council include setting strategic directions, goals, and policies for the funding of health research; approving the CIHR budget; establishing, maintaining, and terminating the Institutes; appointing Scientific Directors as well as IAB members; and providing advice to the Minister of Health.[20]

CIHR is made up of thirteen virtual Institutes.[21] Each Institute has a Scientific Director and an Institute Advisory Board (IAB). The IABs are advisory to the Scientific Directors and to the Governing Council. Each IAB has an Ethics Designate.

CIHR also has a Standing Committee on Ethics. As described by CIHR, the Standing Committee on Ethics identifies for the Governing Council emerging ethical issues of strategic relevance with respect to health and health research. It provides the Governing Council with high-level strategic advice on the ethical, legal, and socio-cultural dimensions of CIHR's mandate as set out in the *CIHR Act*, at its initiative or at the request of the Governing Council or the President.[22] A member of Governing Council is always the Chair of the Standing Committee on Ethics.[23] Surprisingly to some of us, this person need not have any training or expertise in ethics.

17 Canadian Institutes of Health Research Act, s.4(e).

18 Canadian Institutes of Health Research Act, s.4(g).

19 Canadian Institutes of Health Research Act, s.7(1).

20 Canadian Institutes of Health Research, "Governing Council – Role," Government of Canada (January 10, 2014). Available at: www.cihr-irsc.gc.ca/e/38103.html. Accessed on: February 21, 2018.

21 Canadian Institutes of Health Research, "About the Institutes," Government of Canada (August 24, 2016). Available at: www.cihr-irsc.gc.ca/e/7155.html. Accessed on: February 21, 2018.

22 Canadian Institutes of Health Research, "Standing Committee on Ethics," Government of Canada (January 10, 2014). Available at: www.cihr-irsc.gc.ca/e/2859.html. Accessed on: February 21, 2018.

23 Canadian Institutes of Health Research, "Terms of reference," Government of Canada (December 1, 2014). Available at: www.cihr-irsc.gc.ca/e/48875.html. Accessed on: February 21, 2018.

Concerns about the Pfizer Appointment

Our overarching concern with the appointment of a Pfizer executive to the CIHR Governing Council centered on the unmistakable and unavoidable conflict of interest. As one of us (Downie) was quoted in the media at the time:

> This person is in an intractable conflict of interest – on the one hand, he has an obligation to serve the shareholders of his company (as an executive at Pfizer), and on the other hand he would have an obligation to serve the public interest (as a member of the CIHR governing council). Given the divergence of interests between the shareholders and the public, he cannot serve these two masters.[24]

Our first concern with this appointment was with the significant potential for discordance between the interests of Big Pharma and the public. Consider, for example, the CIHR Governing Council's responsibility to establish, maintain, and terminate Institutes.[25] What if an Institute wanted to invest the bulk of its budget in research that would necessarily conflict with the interests of Big Pharma? For example, the Institute of Health Services and Policy Research might want to prioritize funding opportunities for research that (1) investigates how dangerous blockbuster drugs make it to market, (2) examines the problems with ghostwriting by the pharmaceutical industry, or (3) questions the Canadian patent regime with a view to making the production of cheaper generic drugs easier. Clearly, such research priorities could threaten the interests of Big Pharma. In turn, this might influence decision-making about maintaining or terminating this Institute. Consider also the CIHR Governing Council's responsibility to set strategic directions for health research funding. Promoting university-industry partnerships as a funding norm clearly serves the interests of Big Pharma. It is much less clear how this norm equally serves the public interest given that many important research questions are of little interest to the pharmaceutical industry because the answers to these questions cannot easily be

[24] A. Sharratt, "Appointment of Pfizer exec to health funding body criticized," *CBC News* (November 27, 2009). Available at: www.cbc.ca/news/technology/appointment-of-pfizer-exec-to-health-funding-body-criticized-1.789219. Accessed on: February 21, 2018.

[25] Canadian Institutes of Health Research Act, s.20(1)(a).

monetized. An example of such research might be harm reduction strategies for street-involved youth.[26]

In response to concerns about conflict of interest, some suggested that "Prigent's appointment may not represent a direct conflict of interest."[27] From our perspective, however (as well as that of others),[28] the conflict of interest was unmistakably direct. As noted at the time, Prigent, in his role as Vice-President and Medical Director of Pfizer Canada, was under a legal obligation to maximize Pfizer profits. This could, for example, compel him to privilege clinical research over research into societal and cultural dimensions of health and environmental influences on health. As noted above, it could, for example, compel him to privilege research with industry partners over research with community nonprofit partners. It is certainly true that this privileging would not come in the form of direct influence over peer review (i.e., decision-making about which research proposals to fund). However, it could come in the far more pervasively powerful form of direct influence over governance decisions (i.e., setting strategic directions, goals, and policies for health research in Canada).

In addition, some of us noted that the appointment of a senior executive from Pfizer in particular threatened the reputation of CIHR.[29] Just weeks before the appointment, the US Justice Department had issued a press release announcing that the "American pharmaceutical giant Pfizer Inc. and its subsidiary Pharmacia & Upjohn Company Inc. (hereinafter together "Pfizer") [had] agreed to pay $2.3 billion ... to resolve

[26] For a full discussion of the implications of promoting university-industry partnerships as a funding norm, see J. Downie and M. Herder, "Reflections on the commercialization of research," 23–24.

[27] A. Silversides, "CIHR appointment"; A. Silversides, "Appointment of Pfizer executive," p. E256.

[28] S. Lewis, "Neoliberalism, conflict of interest, and the governance of health research in Canada," *Open Medicine*, 1:1 (2010), E28–E30.

[29] For example, see testimony by J. Downie: Parliament of Canada, "Evidence," Standing Committee on Health, Number 049, 2nd Session, 40th Parliament (December 7, 2009). Available at: www.parl.gc.ca/HousePublications/Publication.aspx?DocId=4302528. Accessed on: February 21, 2018. See also J. Downie, "A conflict by any other name would smell as foul: A comment on the appointment of a vice-president of Pfizer to the CIHR Governing Council," *Ottawa Life Magazine* (July 12, 2010). Available at: https://papers.ssrn.com/sol3/papers.cfm?abstract_id=2101686. Accessed on: February 19, 2018.

criminal and civil liability arising from the illegal promotion of certain pharmaceutical products."[30] At the time, this was the largest settlement for health care fraud in the history of the US Justice Department.[31] The $2.3 billion payout included a $1.195 billion criminal fine that, at the time, was also the largest criminal fine in the history of the United States.[32] The appointment of Prigent to the CIHR Governing Council could too easily be seen as CIHR participating in a Pfizer strategy to rehabilitate its reputation.

Given these concerns, we set out to "right the ship." First we tried to engage and mobilize senior leadership within CIHR, ethics people within CIHR, the ethics community outside CIHR, and the Canadian public. Then we tried to directly engage with the federal government through a series of initiatives aided by various Members of Parliament. Sadly, this was a spectacularly unsuccessful set of interventions.

Mobilizing Senior Leadership within CIHR

As noted above, at the time of all of these events, CIHR had thirteen research Institutes, each led by a Scientific Director and an Institute Advisory Board. We shared our concerns about the Pfizer appointment with two CIHR Scientific Directors. While they encouraged us to pursue the matter, they were not prepared to say anything publicly in support of our concerns about conflict of interest (we believe for fear of retribution against them personally or against their Institute). If any of the Scientific Directors held private conversations with Beaudet, then none was successful in persuading him to seek a withdrawal of the government appointment.

We also shared our concerns directly with Beaudet. For example, one of us (Baylis) wrote to Beaudet noting concerns "with both the 'category' for the appointment (i.e., the pharmaceutical industry) and

[30] Quoted in *Mary K. Jones* v. *Pfizer Inc., et al.*, US District Court, Southern District of New York, Case 1:10-cv-03864-AKH. Available at: www.pfizersecuritiessettlement.com/docket-files/993432_1.PDF. Accessed on: February 21, 2018.

[31] *Mary K. Jones* v. *Pfizer Inc., et al.*, US District Court.

[32] *Mary K. Jones* v. *Pfizer Inc., et al.*, US District Court.

the 'company' chosen to represent this category (i.e., Pfizer)."[33] In response, Beaudet vigorously defended the appointment. In so doing, he referenced the federal government's Science and Technology Strategy and noted that "as government fills vacancies on the councils' governing bodies, it will seek out more business and community representation to ensure that the composition of granting council governing bodies reflects Canada's broad economic and national interests."[34]

Beaudet also made selective reference to the CIHR means of realizing the objective through "accelerating the discovery of cures and treatments" and "facilitating the commercialization of health research in Canada and promoting economic development through health research in Canada" (as outlined in the *CIHR Act*), with nary a mention of the means of "promoting, assisting and undertaking research that meets the highest international scientific standards of excellence and ethics," "fostering the discussion of ethical issues and the application of ethical principles to health research," and CIHR's power to "monitor, analyze and evaluate issues, including ethical issues, pertaining to health or health research."[35] Beaudet also provided a description of Prigent's strengths and reputation in relation to health research and insisted that Prigent had been "appointed as an individual and not a representative of Pfizer for his skills, experience, including international clinical experience and personal competencies"[36] – a claim obviously in tension with the previous claim invoking the government's Science and Technology Strategy and the desire to increase business representation.

Finally, Beaudet underlined the fact that Prigent would be subject to the same conflicts of interest guidelines as apply to all CIHR Governing Council members.[37] While this is as it should be, were Prigent to act on the ethical obligation to manage the conflicts of interest by recusing himself from policy decisions that intersected with those of the pharmaceutical industry, then he would not be able to bring to the CIHR

[33] Correspondence from F. Baylis to A. Beaudet (November 5, 2009). On file with the authors.

[34] Correspondence from A. Beaudet to F. Baylis (November 12, 2009). On file with the authors.

[35] Canadian Institutes of Health Research Act, s.5(d).

[36] Correspondence from A. Beaudet to F. Baylis.

[37] Correspondence from A. Beaudet to F. Baylis.

Governing Council the expertise in innovation and commercialization for which he was allegedly appointed. In the very moments when this expertise would be most needed, he would have to recuse himself due to conflict of interest.

Far from entertaining the notion of seeking a withdrawal of the Pfizer appointment, or demonstrating any appreciation of why anyone would be concerned about the appointment, Beaudet insisted that "CIHR was very pleased with the appointment of this exceptional physician to our Governing Council."[38] Others who wrote to Beaudet to express concerns similar to ours were similarly rebuffed.[39] As such, the strategy of marshaling arguments and presenting these to the CIHR Institute Scientific Directors and to the President of CIHR was completely ineffective relative to the goal of protecting and promoting the integrity of our national health research funding agency.

Mobilizing Ethics People within CIHR

As noted earlier, among the members of the CIHR Governing Council, there was a person designated as the ethics representative. That person chaired or cochaired the CIHR Standing Committee on Ethics, which was comprised of eleven voting members including the Chair. As well, at the time of all of these events, each of the thirteen IABs had among its members an Institute Advisory Board Ethics Designate (IAB Ethics Designate).

At the time of the Pfizer appointment, one of us (Baylis) was a CIHR IAB Ethics Designate. Her concerns about the Pfizer appointment were shared with other IAB Ethics Designates at a face-to-face meeting on November 9, 2009. As a group, the IAB Ethics Designates then took their concerns to the CIHR Standing Committee on Ethics at a joint face-to-face meeting later that day. Members of the Standing Committee on Ethics expressed support for these concerns. At the end of the meeting, it was decided that the Committee Chair, Harvey Chochinov (also a member of the CIHR Governing Council), should inform the Governing Council of the IAB Ethics Designates' and the Standing

[38] Correspondence from A. Beaudet to F. Baylis.

[39] Individual correspondences from W. Glannon (November 28, 2009), M. Herder (November 21, 2009), S. Lewis (November 11, 2009), J. Lexchin (November 12, 2009), A. Lippman (November 12, 2009) and A. Schafer (November 11, 2009) to A. Beaudet. On file with the authors.

Committee on Ethics members' shared concerns with the Pfizer appointment. As the Chair had had to leave the meeting early for personal reasons, this decision was conveyed to him by the CIHR staff at the meeting.

The next CIHR Governing Council meeting was ten days later on November 18–19, 2009. The minutes for this meeting do not include a discussion of the Pfizer appointment. The minutes extend a warm welcome to Prigent, "the newest member of Governing Council," and note that the Governing Council "agreed to discuss issues raised in the media and elsewhere related to this appointment at the in camera session."[40] The minutes also include the following opaque statement: "For purposes of information, Governing Council was presented with highlights of the 30th meeting of the Standing Committee on Ethics (SCE), held November 9–10, 2009."[41] As such, there is no formal public record confirming that the CIHR Governing Council was informed of the nature and scope of the ethics concerns raised by the IAB Ethics Designates and the members of the Standing Committee on Ethics. Moreover, correspondence from Chochinov to the Standing Committee on Ethics[42] and testimony by a member of the CIHR Governing Council to the Parliamentary Standing Committee on Health[43] suggest that the Governing Council may not have been given a complete picture of the concerns.

In a letter dated November 26, 2009, Chochinov reported to the CIHR Standing Committee on Ethics that he had understood that "the SCE was uneasy, and wished me [Chochinov] to convey the implications and nuances of this appointment to Governing Council."[44] He assured Committee members that "everyone – including CIHR Senior Executive and the Governing Council – is aware of the issues and has discussed them thoroughly."[45] There is no mention of the fact that it was the IAB Ethics

[40] Canadian Institutes of Health Research, Minutes for 64th Meeting of CIHR Governing Council (November 18–19, 2009). On file with the authors.

[41] Canadian Institutes of Health Research, Minutes for 64th Meeting of CIHR Governing Council.

[42] Correspondence from H.M. Chochinov to the CIHR Standing Committee on Ethics (November 26, 2009). On file with the authors.

[43] J. Rouleau: Parliament of Canada, "Evidence," Standing Committee on Health, Number 049, 2nd Session, 40th Parliament (December 7, 2009). Available at: www.parl.gc.ca/House Publications/Publication.aspx?DocId=4302528. Accessed on: February 21, 2018.

[44] Correspondence from H.M. Chochinov to the CIHR Standing Committee on Ethics.

[45] Correspondence from H.M. Chochinov to the CIHR Standing Committee on Ethics.

Designates who brought these concerns to the Standing Committee on Ethics. And, on December 7, 2009, Jean Rouleau, a member of the CIHR Governing Council, testified before the Parliamentary Standing Committee on Health that he was unaware of the fact that the IAB Ethics Designates had serious concerns about the appointment.[46] He only knew of concerns raised by the Governing Council's Standing Committee on Ethics.[47]

On December 1, 2009, having learned that the CIHR Governing Council had unanimously endorsed the Pfizer appointment, Baylis wrote to Chochinov questioning his apparent decision to set aside the concerns of the IAB Ethics Designates and the members of the Standing Committee on Ethics.[48] In response, Chochinov wrote: "Members of the ethics committee have taught over the years that it is possible for people to reach different conclusions, in spite of following ethically sound processes. I believe the process of ethical debate in this instance has been sound, in spite not reaching consensus [*sic*]."[49] In addition, Chochinov provided Baylis with a copy of his November 26, 2009, letter to the CIHR Standing Committee on Ethics. In this letter Chochinov offered five reasons for having supported the Pfizer appointment: (1) Prigent was not there to promote his own vested interests or those of his company; (2) Prigent's expertise would help CIHR with its commercialization and economic development mandate (with no mention of how this appointment might hurt the CIHR ethics mandate); (3) all members of Governing Council have to observe the Conflict of Interest Act, the Ethical Guidelines for Public Office Holders, and the Guidelines for the Political Activities of Public Office Holders; (4) Prigent was a leader in the promotion of Research and Development within the Canadian Life Science Sector; and (5) Prigent would only be one voice among many with richly varied backgrounds and so his appointment would not undermine the independence of research at CIHR.[50]

[46] J. Rouleau: Parliament of Canada, "Evidence."
[47] J. Rouleau: Parliament of Canada, "Evidence."
[48] Correspondence from F. Baylis to H.M. Chochinov (December 1, 2009). On file with the authors.
[49] Correspondence from H.M. Chochinov to F. Baylis (no date). On file with the authors.
[50] Correspondence from H.M. Chochinov to the CIHR Standing Committee on Ethics. Correspondence from Members of CIHR Governing Council (K. Anderson, J. Brien, H.M. Chochinov, and B.B. Finlay) to Judy Wasylycia-Leis (November 30, 2009). On file with the authors.

In testimony presented to the Parliamentary Committee on Health, much ado was made of the fact that all members of the CIHR Governing Council had endorsed the Pfizer appointment.[51] There was no mention of the fact that contrary to the current ethics member of CIHR Governing Council (who lacked formal training in ethics), the two previous ethics members of CIHR Governing Council – Françoise Baylis and Nuala Kenny, both of whom had formal training in ethics – had issued a press release stating unequivocally that, had they still been the ethics member of Governing Council, they would not have endorsed this appointment, in which case there would not have been unanimous support for the Pfizer appointment.[52]

Our strategy of presenting the CIHR Governing Council with evidence of unanimous ethics concern by the IAB Ethics Designates and the Standing Committee on Ethics (all but the Chair) proved to be entirely unsuccessful.

Mobilizing the Ethics Community Outside CIHR and the Public

On November 24, 2009, with the help of our colleague Matthew Herder, we took our ethics concerns to a wider audience by inviting Canadians to sign an online petition at Go Petition.[53] Another colleague, Marc-André Gagnon, provided a French translation. The bilingual petition calling for the withdrawal of the Pfizer appointment was directed at the House of Commons. We emailed the URL for the petition to colleagues, family, and friends in an effort to drive traffic to the online petition (at the time, there were not the myriad petition sites available today and neither of us was on Facebook, Twitter, or other social media – although tech-savvy friends and family spread the word for us).

[51] A. Beaudet: Parliament of Canada, "Evidence," Standing Committee on Health, Number 047, 2nd Session, 40th Parliament (November 30, 2009). Available at: http://www.ourcommons.ca/DocumentViewer/en/40-2/HESA/meeting-47/evidence. Accessed on: February 21, 2018.

[52] Faculty of Medicine Media Advisory, "Dal bioethicists condemn pharma exec's appointment to Canadian health research institute," Dalhousie University (December 7, 2009). On file with the authors. See also J. Downie: Parliament of Canada, "Evidence."

[53] M. Herder (organizer), "Petition against the appointment of the VP of Pfizer Canada to CIHR Governing Council/Pétition contre la nomination du vice-président de Pfizer Canada au conseil d'administration des IRSC" (November 24, 2009–January 25, 2010). Available at: www.gopetition.com/petition/32371.html. Accessed on: May 13, 2016. No longer publicly accessible; on file with author.

Immediately, the signatures started rolling in – more than a 1,000 in two days. As reported in the press:

> Herder, Downie and Françoise Baylis . . . drafted a petition opposing the appointment on Wednesday. As of Friday, it had 1,320 signatures . . . Wasylycia-Leis [New Democratic Party (NDP) Member of Parliament] says she will present the petition to the committee [Standing Committee on Health] and to the House of Commons, but she's worried that the committee is stacked against her. The chair of the committee has turned down her request to present witnesses who can authoritatively discuss the issues Prigent's appointment raises, she said.[54]

As the number of signatures increased, so did media attention.[55] Downie did an interview on *CBC Information Morning* (broadcast throughout Nova Scotia) during which she alerted citizens to the problem and invited them to sign the petition. There was a notable spike in signatures from Nova Scotia immediately following the broadcast. National media coverage would come later (*CBC The Current*[56] and *CBC News*[57]), including CBC's own poll on the matter.[58] The petition made the "news" in several online media, including websites and blogs dedicated to these issues.[59] This extended the reach of our message well beyond Canadian borders.

54 A. Sharratt, "Appointment of Pfizer exec to health funding body criticized."

55 A. Picard, "The man from Pfizer: Should Big Pharma help steer health research?," *Globe and Mail* (December 3, 2009). Available at: www.theglobeandmail.com/life/health-and-fitness/the-man-from-pfizer-should-big-pharma-help-steer-health-research/article4311392/. Accessed on: February 21, 2018.

56 A. Tremonti, "Pt 1 – CIHR and Pfizer," *CBC The Current* (December 9, 2009). Originally available at: http://castroller.com/podcasts/TheCurrent/1360077. No longer publicly accessible.

57 "Pfizer exec's health-funding job defended," *CBC News – Technology & Science* (December 3, 2009). Available at: www.cbc.ca/news/technology/pfizer-exec-s-health-funding-job-defended-1.848453. Accessed on: February 21, 2018.

58 "Pfizer appointment: Is it a conflict of interest?," *CBC News – Point of View* (December 4, 2009). Available at: www.cbc.ca/news2/pointofview/2009/12/pfizer-appointment-is-it-a-conflict-of-interest.html. Accessed on: February 21, 2018.

59 "Appointment of Pfizer VP to CIHR generates controversy," *Academica Group* (December 10, 2009). Available at: https://www.academica.ca/top-ten/appointment-pfizer-vp-cihr-generates-controversy. Accessed on: February 21, 2018. "Get that Pfizer out of my CIHR," *bayblab* (December 23, 2009). Available at: http://bayblab.blogspot.ca/2009/12/get-that-pfizer-out-of-my-cihr.html. Accessed on: February 21, 2018. T. Bousquet, "Bernard Prigent, Pfizer's inside man," *The Coast* (February 4, 2010). Available at: https://www.thecoast.ca/halifax/bernard-prijent-pfizers-inside-man/Content?oid=1503474. Accessed on: February 21, 2018. Concernedresearchers2, "CIHR appointment of Pfizer exec stirs outrage," *Researcher Forum* (December 7, 2009). Available at: https://dontleavecanadabehind.wordpress.com/2009/12/07/cihr-appointment-of-pfizer-exec-stirs-outrage/. Accessed on:

On December 3, 2009, to let more people know about the petition, Baylis published "An intractable conflict of interest" in *The Mark News* (reprinted by *Longwoods*) laying out our concerns and CIHR's responses to date.[60] When the petition officially closed on January 25, 2010, some 4,411 Canadians (including CIHR-funded researchers, Canada Research Chairs, members of the Royal College of Physicians and Surgeons, members of the Royal Society of Canada and of the Canadian Academy of Health Sciences, Deans of Medicine, Nursing, Health Professions, and Law, and many members of the general public) had signed the petition.[61] And while a petition with this number of signatures might be considered insignificant by today's standards (in an era of Change.org), it was a remarkable number of signatures at the time, especially given the esoteric topic – viz., membership on the Governing Council of Canada's health research funding agency – about which it would be considered difficult to engage members of the general public. One of the many heartening features of the call for public support was the fact that those who responded "got it" – they understood the conflict of interest as evidenced by such comments submitted with signatures as "You don't put the rooster in charge of the hen house."[62]

This effort, aimed at persuading Parliament to withdraw the Pfizer appointment, proved less effective than we had hoped. We were not able to have the petition officially presented to the House of Commons because it did not follow the form and content requirements to be certified and presented to the House[63] (we were unaware of the rules

February 21, 2018. E. Silverman, "A petition to keep Pfizer exec off Canadian board," *Healthy Skepticism*, Library item:16832 (November 25, 2009). Available at: http://www.healthyskepticism.org/global/library/item/16832. Accessed on: February 21, 2018. R. Tamburri, "Federal granting agencies get more appointees from outside academe. Outcry over a Pfizer executive appointed to CIHR's governing body hasn't echoed with the granting councils," *University Affairs/Affaires universitaires* (February 8, 2010). Available at: https://www.universityaffairs.ca/news/news-article/federal-granting-agencies/. Accessed on: February 21, 2018. N. Walton, "Should Pfizer VP sit on CIHR's Governing Council?," *Research Ethics Blog* (November 29, 2009). Available at: https://researchethicsblog.com/2009/11/29/should-pfizer-vp-sit-on-cihrs-governing-council/. Accessed on: February 21, 2018.

60 F. Baylis, "An intractable conflict of interest," *The Mark News* (December 2, 2009). Reprinted in: *Open Medicine* (December 2009). Available at: http://www.longwoods.com/content/21200. Accessed on: February 21, 2018.

61 M. Herder (organizer), "Petition against the appointment of the VP of Pfizer."

62 M. Herder (organizer), "Petition against the appointment of the VP of Pfizer."

63 Parliament of Canada, "Public Petitions," in R. Marleau and C. Montpetit (eds.), *House of Commons Procedure and Practice* (2000). Available at: http://www.ourcommons.ca/MarleauMontpetit/DocumentViewer.aspx?Language=E&Sec=Ch22&Seq=3. Accessed on:

governing the presentation of a petition to the House as we had successfully used this strategy once before in relation to the Assisted Human Reproduction Act 2004, without running afoul of any rules[64]). However, we were able to get the petition before the Parliamentary Standing Committee on Health with some initial positive effect, as described below.

Mobilizing Parliamentarians

Several concerned Canadians tried to engage directly with the Minister of Health, Leona Aglukkaq.[65] This too proved to be an ineffective intervention. In March 2010, the Minister of Health returned a form letter (i.e., a letter with identical content) to several individuals who had written to her independently stipulating that:

> While in office, the conduct and action of CIHR Governing Council appointees is governed by specific statues and guidelines ... It should also be noted that as a member of the Governing Council, Dr. Prigent will only be involved in high level budget decisions. He will not make funding decisions with respect to which researchers or projects would be funded.[66]

This response not only sidestepped, but also effectively misrepresented the concerns of many, which were very clearly about "high level budget decisions" and the ways in which this high-level budgetary decision-making could be distorted (the concerns were not about funding decisions). More effective were the interventions directed to specific members of Parliament who were not part of the government. Two interventions are particularly worth mentioning.

First, Steven Lewis (a health policy consultant, former CEO of a provincial granting agency, and former member of the CIHR Governing Council) wrote to Carolyn Bennett, a Liberal Member of Parliament, after she was quoted in the *Canadian Medical Association Journal* as

February 21, 2018. Parliament of Canada, Standing Orders: Chapter IV, Standing Order 36(1). Available at: www.parl.gc.ca/About/House/StandingOrders/chap4-e.htm. Accessed on: February 21, 2018.

[64] Open Letter from 65 Canadian Health Care Ethics and Health Law Experts on Bill C-13 (October 26, 2003). Press release; distributed to members of Parliament. On file with the authors.

[65] Correspondence from F. Baylis (November 11, 2009), C. Elliott (December 4, 2009), C. Fuller (no date), C. Kaposy (November 17, 2009), T. Lemmens (November 26, 2009) and U. Schuklenk (December 4, 2009) to L. Aglukkaq. On file with the authors.

[66] Correspondence from L. Aglukkaq to F. Baylis (March 19, 2010). On file with the authors.

supporting the Pfizer appointment. He provided her with arguments against the appointment and explained that after consulting “dozens of health researchers and policy analysts” it was found that they were “unanimously concerned by and opposed to this appointment.”[67] He asked her to:

> speak in favour of reversing the appointment, and more importantly in favour of transparent policies that recognize the importance of maintaining a respectful distance between the pursuit of scientific excellence in the public interest, and organizations whose obligation is to private shareholders.[68]

Baylis also wrote to Bennett asking her to reconsider her “apparent support for this appointment”[69] and followed up with a phone call.

At this same time, we contacted our local Member of Parliament, Megan Leslie (a member of the NDP), to learn more about parliamentary processes. Leslie agreed to ask a question in the House of Commons. We then presented her with a draft two-page speech followed by a question. After she stopped laughing – no doubt on account of our naiveté – Leslie told us that, even if she were able to get her caucus to agree to give her one of their few allocated question slots, she would only have thirty-five seconds within which to ask her question. We were obviously a very long way from thirty-five seconds. We agreed to change tack. Instead, Leslie would try to get the issue in front of the Parliamentary Standing Committee on Health, and she would place written questions for the Minister of Health on the *Order Paper*.[70]

In pursuit of the first of these two strategies, Leslie discussed our concerns with Judy Wasylycia-Leis (NDP Member of Parliament and Vice Chair of the Standing Committee on Health). On November 13, 2009, Wasylycia-Leis submitted a notice of motion to the Standing Committee on Health seeking a review of the Order in Council appointment of Prigent to the Governing Council of CIHR.[71] She also issued a press release in which she described the situation with reference to a

[67] Correspondence from S. Lewis to C. Bennett (November 20, 2009). On file with the authors.

[68] Correspondence from S. Lewis to C. Bennett.

[69] Correspondence from F. Baylis to C. Bennett (November 20, 2009). On file with the authors.

[70] See M. Leslie (Halifax), “Questions,” Parliament of Canada, Q-32^2 (March 3, 2010). Available at: www.parl.gc.ca/HousePublications/Publication.aspx?DocId=4320637&Language=E&Mode=2&Parl=40&Ses=3&File=11. Accessed on: February 21, 2018.

[71] J. Wasylycia-Leis, Notice of motion (November 13, 2009). On file with the authors.

popular children's fairy tale, "The Three Little Pigs": "Having drug company executives deciding health research directions is like having the big bad wolf directing the three little pigs on how to build their homes."[72]

Prigent and Beaudet were invited to appear as witnesses before the Standing Committee on Health, as were Trudo Lemmens (a professor from the Faculties of Law and Medicine at the University of Toronto) and Noralou Roos (a professor in the Faculty of Medicine at the University of Manitoba and a past member of the CIHR Interim Governing Council). The invitations to Lemmens and Roos were issued on November 26 by Judy Wasylycia-Leis, but on November 27, both Lemmens and Roos were disinvited by members of the Conservative party controlling the Committee.[73] This meant that the Standing Committee on Health would hear testimony only from those with vested interests in defending the contentious appointment.

At the beginning of the November 30 meeting, Wasylycia-Leis tabled a snapshot of the online petition, noting that it already included "some 2,085 names of concerned citizens" and referenced correspondence from people concerned about the appointment.[74] She also took exception to the fact that witnesses who were opposed to the appointment had not been invited to address the committee. Bennett (who had expressed support for the Pfizer appointment prior to receiving the information from Lewis described earlier) noted that she had received a "tsunami of objections" from "some of the most prominent health ethicists, professors, and health law professors" with "very eloquent and persuasive arguments."[75] She argued that the Committee should hear from at least

72 "Drug V.P. should not sit on health council: New Democrats; Pfizer V.P. appointment is a conflict of interest and cause for concern," press release (November 16, 2009). On file with the authors. Quoted in: A. Silversides, "Federal committee to review Pfizer v-p appointment to CIHR council," *CMAJ News* (early release, November 19, 2009). Available at: www.cmaj.ca/content/early/2009/11/19/cmaj.109-3113. Accessed on: February 21, 2018. In print: A. Silversides, "Federal committee to review appointment of Pfizer vice-president to CIHR governing council," *CMAJ*, 182:1 (2010), E33–E34.

73 Personal communication between Françoise Baylis, Trudo Lemmens, and Noralou Roos, supported by written email correspondence between Lemmens and Roos (November 29, 2009). On file with the authors.

74 J. Wasylycia-Leis: Parliament of Canada, "Evidence," Standing Committee on Health, Number 047, 2nd Session, 40th Parliament (November 30, 2009). Available at: http://www.ourcommons.ca/DocumentViewer/en/40-2/HESA/meeting-47/evidence. Accessed on: February 21, 2018.

75 C. Bennett: Parliament of Canada, "Evidence," Standing Committee on Health, Number 047, 2nd Session, 40th Parliament (November 30, 2009). Available at: http://www

one witness who was opposed to the appointment. After some discussion, the Committee agreed to give a second one hour of discussion time to the issue at the next Committee meeting.

Following this procedural scuffle, the Committee heard from Prigent and Beaudet. In his testimony, Prigent described his past research and business experience and in closing stated, "I undertake to strictly observe the Conflict of Interest Act, the Ethical Guidelines for Public Office Holders, the Guidelines for the Political Activities of Public Office Holders and the CIHR's Policy on Conflict of Interest and Confidentiality in the Context of Merit, Relevance and Peer Review."[76] Later in response to a question from Wasylycia-Leis, Prigent added, "If there's one thing I pay attention to, it's the conflict of interest. I think it's important."[77] Beaudet, for his part, insisted that Prigent was not representing Pfizer, but rather was representing himself. He also insisted on the importance of working together and aligning the CIHR and pharmaceutical industry "agendas" and "vision."[78] Beaudet also revealed that he was the person who had submitted Prigent's name to the Minister of Health. To this point, many of us had thought that the Pfizer appointment had been forced upon CIHR by the government. We were stunned to find out otherwise.

During the question period, Bennett raised concerns about conflicts of interest (both real and apparent[79]) that would be faced by Prigent. When she tried to raise the matter of Pfizer's record and the recent $2.3 billion fine, she was ruled out of order by the Chair. The scope of the meeting was said to relate directly to Prigent's ability to perform the job, not any egregious behavior in which his company may or may not have been involved.[80] Despite her being shut down, it is clear that Lewis's and Baylis's

.ourcommons.ca/DocumentViewer/en/40-2/HESA/meeting-47/evidence. Accessed on: February 21, 2018.

76 B. Prigent: Parliament of Canada, "Evidence," Standing Committee on Health, Number 047, 2nd Session, 40th Parliament (November 30, 2009). Available at: http://www.ourcommons.ca/DocumentViewer/en/40-2/HESA/meeting-47/evidence. Accessed on: February 21, 2018.

77 B. Prigent: Parliament of Canada, "Evidence."

78 A. Beaudet: Parliament of Canada, "Evidence."

79 For a discussion of this distinction as applied to the appointment of Dr. Prigent to the Governing Council of CIHR, see E. Gibson, "Concepts of bias and appointments to the Governing Council of the Canadian Institutes of Health Research," *CMAJ*, 182:17 (2010), E793–E795.

80 Parliament of Canada, "Evidence," Standing Committee on Health, Number 047, 2nd Session, 40th Parliament (November 30, 2009).

interventions had been effective. In our opinion, Bennett exhibited something admirable yet seemingly quite rare among politicians – namely, a willingness to change her position on an issue (and acknowledge having done so) when presented with compelling new information.

The following week, on December 7, 2009, more witnesses appeared before the Standing Committee on Health, including Jocelyn Downie, Trudo Lemmens, Jean Rouleau, and Peter Brenders. While this second one-hour session before the Committee was scheduled so that persons opposed to the Pfizer appointment could explain their concerns, the witnesses were evenly divided between those endorsing and those opposing the appointment. Downie and Lemmens, both professors in Faculties of Law and Medicine, were there to explain concerns about conflict of interest, whereas Rouleau (then a sitting member of the CIHR Governing Council and Dean of the Faculty of Medicine at the University of Montreal) and Brenders (a registered lobbyist for the biotech industry and member of BIOTECanada) were there to speak in support of the Pfizer appointment.

Having learned from the response to Bennett during the question period in the November 30 meeting, Downie was particularly careful during the December 7 meeting to frame all of her comments in terms of the qualifications of the appointee.[81] This was essential in order to avoid being ruled out of order by the Chair. By approaching her testimony in this way, Downie was able to get Prigent's conflicts of interest and Pfizer's record of misconduct in the official record.[82]

In their testimony before the Standing Committee on Health, Beaudet, Brenders, and Rouleau independently claimed that Prigent likely would not face irreconcilable conflicts of interest in his dual role as Vice President of Pfizer Canada and member of the CIHR's Governing Council.[83] They further insisted that if an actual or apparent conflict of interest were to arise, Prigent would recognize this and recuse himself. In rather colorful language, Rouleau testified that it was not as if

> the other 15 members of GC [Governing Council] and the leadership of CIHR just fell off a turnip truck and would be unable to recognize an

[81] J. Downie: Parliament of Canada, "Evidence."
[82] J. Downie: Parliament of Canada, "Evidence."
[83] A. Beaudet, P. Brenders, and J. Rouleau: Parliament of Canada, "Evidence," Standing Committee on Health, Number 049, 2nd Session, 40th Parliament (December 7, 2009). Available at: http://www.ourcommons.ca/DocumentViewer/en/40-2/HESA/meeting-49/evidence. Accessed on February 21, 2018.

> intervention that is coloured by conflict of interest . . . the members of GC are nearly all members of the institutions that receive funding from the CIHR. We are all aware of this potential conflict of interest . . . and hold one another to the highest ethical standards possible, as we will with Dr. Prigent.[84]

Meanwhile, undermining their claims, none of those testifying in defense of the Pfizer appointment (Beaudet, Brenders, or Rouleau) proactively disclosed their own significant and relevant conflicts of interest to the Standing Committee on Health.[85] At the time, Brenders was CEO of BIOTECanada, and Beaudet and the President and CEO of Pfizer were on the Board of Directors of BIOTECanada.[86] Brenders was a member of the "Friends of CIHR."[87] Brenders was a registered lobbyist for the biotech industry.[88] Rouleau's significant connections with Pfizer included donations from Pfizer to the Faculty of Medicine of which he was Dean,[89] research funding from Pfizer, coauthoring with Pfizer staff and service on Pfizer speakers' bureau, and consulting for Pfizer.[90] In November 2009, CIHR gave Pfizer the Prix Galien Innovative Product Award ("the most prestigious award in the field of Canadian pharmaceutical research and

[84] J. Rouleau: Parliament of Canada, "Evidence."

[85] J. Rouleau: Parliament of Canada, "Evidence."

[86] "BIOTECanada announces appointments to Board of Directors for 2009–10," *Business Wire*. Available at: http://www.businesswire.com/news/home/20090508005330/en/BIOTECanada-announces-appointments-Board-Directors-2009-10. Accessed on: February 21, 2018.

[87] Parliament of Canada, "Evidence," Standing Committee on Health, Number 049, 2nd Session, 40th Parliament (December 7, 2009). Available at: http://www.ourcommons.ca/DocumentViewer/en/40-2/HESA/meeting-49/evidence. Accessed on: February 21, 2018.

[88] Parliament of Canada, "Evidence."

[89] "Université de Montréal creates Pfizer Chair in Atherosclerosis Leading atherosclerosis researcher Dr. Jean-Claude Tardif is first incumbent. A $1.8 million partnership in support of innovative research into the world's leading cause of death," i-Forum Communiqué, Université de Montréal Archives des communiqués (February 3, 2004). Available at: http://www.iforum.umontreal.ca/Communiques/ArchivesCommuniques/2004/3173.htm. Accessed on: February 21, 2018.

[90] J.L. Rouleau, W.J. Warnica, R. Baillot, P.J. Block, S. Chocron, D. Johnstone, et al., "Effects of angiotensin-converting enzyme inhibition in low-risk patients early after coronary artery bypass surgery," *Circulation*, 117:1 (January 2008), 24–31. P.W. Armstrong and J.L. Rouleau, "A Canadian context for the acute study of clinical effectiveness of Nesiritide and decompensated heart failure (ASCEND-HF) trial," *Canadian Journal of Cardiology*, 24:Suppl B (July 2008), 30B–32B.

innovation") for Champix.[91] Some, but not all, of this information surfaced in the meeting of the Standing Committee on Health, but only in response to direct questions from Judy Wasylycia-Leis.[92] Meanwhile, Rouleau assured the Committee that "we [GC members] are watching each other very closely to ensure that everyone adheres to the ethics code" and that "we are watching each other to ensure the best ethical practices possible."[93]

Following this second meeting, the Standing Committee on Health concluded that it would not call for the withdrawal of the Pfizer appointment, but that it would discuss how to follow up on the issues raised by the Pfizer appointment at their next meeting in January 2010.[94] Some members of the Committee were keen to call back the President of CIHR as well as ethics experts for further discussions about how the conflicts of interest identified during the two hearings were actually being managed in the functioning of the CIHR Governing Council. We have been unable to find evidence of such testimony being elicited or such discussions ever being held.[95]

Lessons Learned

A number of lessons can be drawn from this experience. First, ethics training and expertise matters. For example, had the ethics member of the Governing Council had such training and expertise, there is reason to believe that there would not have been unanimous support on the part of Governing Council for introducing such an egregious conflict of interest into the CIHR governance structure. Second, a petition signed by a large number of Canadians (including eminent Canadians) gets media attention. This was very important in getting the attention of politicians. Without the online petition, the

[91] "Government of Canada pays tribute to the country's top health researchers," *Prix Galien Canada*, press release (November 17, 2009). Available at: http://eng.prix-galien-canada.com/press_2009_Awards.html. Accessed on: February 21, 2018.

[92] Parliament of Canada, "Evidence." [93] J. Rouleau: Parliament of Canada, "Evidence."

[94] J. Wasylycia-Leis: Parliament of Canada, "Evidence," Standing Committee on Health, Number 049, 2nd Session, 40th Parliament (December 7, 2009). Available at: http://www.ourcommons.ca/DocumentViewer/en/40-2/HESA/meeting-49/evidence. Accessed on February 21, 2018.

[95] Our searches included documents collected under Parliament of Canada, "Meetings," Standing Committee on Health. Available at: www.parl.gc.ca/Committees/en/HESA/Meetings. Accessed on: February 21, 2018.

Parliamentary Committee might never have held hearings on the issue, thereby creating a very public record of the ethics disagreement. Third, great care must be taken in testifying before a Parliamentary Committee to ensure that comments (and answers to questions) do not stray from the scope of a Committee's inquiry. Otherwise, no matter how important the comments, they may easily be ruled out of order by the Chair (a member of the governing party). Fourth, when a majority federal government and one of its (arm's-length) agencies have determined a course of action, it can be frustratingly impossible to convince them to change course, notwithstanding well-informed credible arguments against the chosen course of action and notwithstanding pressure from internal ethics experts; external ethics, law, policy, and research experts; elected officials who are not members of the governing party; and the general public. When that action aligns with the interests of Big Pharma, we all need to be wary. Finally, in taking on a Sisyphean task (as this one turned out to be), it is always wise to do so in the company of others. This helps not only to spread the work around, but also to keep up the morale in the face of repeated setbacks and defeats. We could not have done this work over several months without the support of each other and the concrete help of untold members of the bioethics community.

Postmortem

Despite all of our efforts to challenge Prigent's appointment, not only did he retain his seat at the CIHR Governing Council, but his initial three-year term appointment was renewed for a second term. In total, Prigent served from 2009 to 2015. Furthermore, a careful review of the CIHR Governing Council Minutes for the six years that Prigent was a member of the Governing Council reveals that Prigent declared a conflict of interest only once – he stepped out of the room during a discussion of appointments to Governing Council Committees through which he was appointed to the Audit Committee.[96]

But happily, this is not entirely a story of failure. When Prigent finished his two terms on Governing Council, he was not replaced by someone from Pharma (big or small). The CIHR Governing Council counts among its

[96] Canadian Institutes of Health Research, Minutes for the 86th Meeting of CIHR Governing Council (November 18, 2013), p. 3. On file with the authors.

members individuals with business acumen and business affiliations, but none is an employee of a pharmaceutical company.[97] Long may that last.

[97] That said, it must be noted that Jennifer Hamilton, Senior Director, New Ventures (Canada), Johnson & Johnson is a member of the CIHR Standing Committee on Ethics. Available at: www.cihr-irsc.gc.ca/e/50242.html. Accessed on: February 21, 2018.

3

"So What?"

Historical Contingency, Activism, and Reflections on the Studies in Tuskegee and Guatemala

SUSAN M. REVERBY*

I am a baby boomer, a child of the 1960s who lost her political innocence in the bloodshed of the Vietnam War era, yet cemented a sense of agency and civic responsibility in the civil rights, antiwar, women's, and health movements. Even as I became an academic, I retained my activist roots that grew my sense that confronting immoral power and finding collective solutions were a life's work. I spent three formative years in the early 1970s, between dropping out and continuing graduate school, as a radical health politics journalist in a New York think tank called the Health Policy Advisory Center (Health/PAC). Considered the intellectual center of the burgeoning health activist movement, at Health/PAC I learned to write quickly, debate those in power when I still knew very little, support various kinds of actions against what we labeled the "medical health empires," and understand the structure of the health care system.[1]

I went back to graduate school in American studies/history because I really did believe in footnotes and facts and that discovering and using primary documents to make ordered arguments about the past could guide the present and perhaps even the future. In this regard, I share the views of historian/health administrator Daniel M. Fox: "Beliefs about history have had an enormous influence on health policy in the United States and abroad. Historians may have a subordinate role in public

* I wish to thank Alice Dreger and Françoise Baylis for the opportunity to reflect on my experiences, and their comments on an earlier draft.

[1] See M. Chowkwanyun, "The new left and public health: The Health Policy Advisory Center, community organizing, and the big business of health, 1967–1975," *American Journal of Public Health*, 101 (February 2011), 238–249.

affairs, but historicism – the notion that history teaches, that the past is prologue – retains enormous power."[2]

As I began my academic career, one of my closest friends taught me her two most important words for evaluating research: "So what?" That is, does your scholarship really produce important new knowledge or analysis?[3] And for me there were additional issues: Was I trying to be a scholar whose work attempted to reach a broader audience, acknowledged a social problem, and might even point a way to social change? This chapter is an effort to explain how these concerns affected how I became drawn to researching and writing about two of the biggest unethical government-funded research studies in American history: the US Public Health Service Study in Tuskegee, Alabama, and the Sexually Transmitted Diseases Inoculation Studies in Guatemala. A scholar can pick a particular topic, and hope that the answer to the "so what" of relevant questions and concerns will be obvious, or teased out in her work. But it is often catching a particular moment, historical contingency, and collective efforts that take the "so what" beyond the academy to social change. Just like restorative justice where there is an effort to involve the perpetrator(s), the victim(s), and their communities, there can be restorative history that does the same.[4] Yet the historical has to become consciously political if it is to really make a difference.

Tuskegee as History, Metaphor, and Memory

The bare outlines of the study in Tuskegee are widely known, especially in health research and bioethics communities, since the details are so appalling and have been crucial to the growth of research

[2] D.M. Fox, "History and health policy: An autobiographical note on the decline of historicism," *Journal of Social History*, 18 (Spring 1985), 349–364, at 352. For my earlier discussion of this, see S.M. Reverby, "Thinking through the body and the body politic," in G. Feldberg, M. Ladd-Taylor, A. Li, and K. McPherson (eds.), *Women, Health and Nation: Canada and the United States since 1945* (Montreal: McGill-Queens University Press, 2003), pp. 404–420.

[3] Thanks to my Wellesley College Women's and Gender Studies colleague Rosanna Hertz for this terminology. She recalls that her Northwestern PhD advisor, Allan Schnaiberg, asked her this kind of question, but not in these words. See also A. Kessler-Harris, "The so-what question," *Frontiers*, 36 (2015), 12–16.

[4] S.M. Reverby, "Restorative justice and restorative history for the sexually transmitted disease inoculation experiments in Guatemala," *American Journal of Public Health*, 106 (July 2016), e1–e2.

regulation and ethics training.[5] It became publicly known nearly fifty years ago and has had the time to become embedded in cultural consciousness.[6] The "Tuskegee Study of Untreated Syphilis in the Negro Male," the research's published title, was a forty-year endeavor (1932–1972) on the part of the US Public Health Service to study, but not treat, hundreds of African American men with late stage and presumed noninfectious syphilis, while promising the men that the aspirins, tonics, and diagnostic spinal taps provided were treatment.[7] Coming at the end of the Civil Rights era and building on centuries of justifiable fears over the role of medicine in the perpetuation of racism, information and misinformation about the study circulated in popular and political culture after 1972.[8] Yet the term "Tuskegee," when used in the context of medical care and clinical research has come to mean so much more than just this study.[9]

Today, "Tuskegee" is a powerful metaphor for the refusal to consider African Americans as rights-bearing citizens – almost as potent as the facts about slavery, lynching, and police killings. The study's shorthand name "Tuskegee" is used to describe behaviors and events ranging from racism in health care, denial of services, and government malfeasance to cultural mistrust as distinct from Tuskegee University.[10] It is called upon in settings as diverse as a *Saturday Night Live* comedic routine about race and medicine, judicial decisions on other immoral experiments, and the efforts of the political right to drum up African American

[5] "How Tuskegee Changed Research Practices," Centers for Disease Control and Prevention (December 14, 2015). Available at: www.cdc.gov/tuskegee/after.htm. Accessed on: October 12, 2016.

[6] J. Heller, "Syphilis victims in U.S. study went untreated for 40 years," *New York Times* (July 26, 1972), p. 1; S.M. Reverby, *Examining Tuskegee: The Infamous Syphilis Study and Its Legacy* (Chapel Hill, NC: University of North Carolina Press, 2009).

[7] The study was primarily known as "Untreated Syphilis in the Male Negro" until a medical report in 1954 called it the "Tuskegee study," named after the location in Tuskegee, Alabama. In 1972, when the first newspaper account of the story broke, the term "Tuskegee Study" or "Tuskegee Experiment" became its more public name. During the organizing for a federal apology obtained from President Bill Clinton in 1997, the effort was made to begin to call it the "United States Public Health Service Syphilis Study." For more on the difficulties of what to call it, see S.M. Reverby, *Examining Tuskegee*, pp. 1–12.

[8] For why the Tuskegee Study is not the only issue, see V.N. Gamble, "Under the shadow of Tuskegee: African Americans and health care," in S.M. Reverby (ed.), *Tuskegee's Truths: Rethinking the Tuskegee Syphilis Study* (Chapel Hill, NC: University of North Carolina Press, 2000), pp. 431–442.

[9] S.M. Reverby, *Examining Tuskegee.*

[10] S.M. Reverby, *Examining Tuskegee.*

community support to oppose state funding for stem cell research.[11] Indeed, "Tuskegee" has taken on mythic proportions, available to be "use[d] and abuse[d]."[12]

In part, this re-remembering/misremembering of "Tuskegee" happens because the wounds of the study and the myriad other experiences of African Americans with health care have not been healed by the resulting lawsuit, Senate hearings, a federal investigating committee, authoritative historical accountings, or even finally a federal apology in 1997.[13] "Tuskegee" is not merely a metaphor or symbol, then, but rather a prime example of the exploitation of the vulnerable, often racialized minorities, in the name of medical science.[14]

When I started to think about doing research on the study in Tuskegee in the early 1990s, I had just completed a book on the history of American nursing.[15] I thought I might write about Eunice Rivers Laurie, the African American nurse who was the go-between hired by the Public Health Service and Tuskegee Institute to be almost a private public health nurse for the men in the study and their families. I began to collect the available primary and secondary materials. I went to Tuskegee to do research in the university's archives, interviewed people in the community, and then visited various other archives across the country that held materials about the study.

I began to give a number of talks to nursing and history audiences on Nurse Rivers, as she was known most of her life. I tried to contextualize her difficulties, and even to argue that she may have worked covertly to get some of the men to treatment. I spent a lot of time trying to correct the widespread false assumption that the men in the study in Tuskegee had been *given* syphilis by the government.[16] In these talks, I explained that the men already had syphilis, that what the government did was lie to them about the nature of the interventions they were providing. The

11 S.M. Reverby, *Examining Tuskegee*.

12 A. Fairchild and R. Bayer, "Uses and abuses of Tuskegee," in S.M. Reverby (ed.), *Tuskegee's Truths*, pp. 589–604.

13 S.M. Reverby (ed.), *Tuskegee's Truths*.

14 R.V. Katz and R. Warren (eds.), *The Search for the Legacy of the USPHS Syphilis Study at Tuskegee* (Lexington, MA: Lexington Books, 2011).

15 S.M. Reverby, *Ordered to Care: The Dilemma of American Nursing* (New York, NY: Cambridge University Press, 1987).

16 The belief that the men were given syphilis circulates everywhere: from NBC nightly news broadcasts to scholarly papers to conspiracy theories. The occasions on which it occurs are too numerous to list, but I have spent the better part of two decades trying to correct this wherever and whenever I can.

men were supposed to be kept from active treatment that was otherwise available, although I argued that by the antibiotic era there were many examples of their finding their way to treatment. While I think the talks I gave and the subsequent article I wrote helped to explain the complexity of Nurse Rivers's liminality, I wanted to make this into a bigger story – one in which she would not have to carry the weight of the story of the study in Tuskegee, which seemed to me to be immensely unfair.[17]

I think I knew there were not enough primary materials for a book on Nurse Rivers, but I went ahead and wrote a book proposal. I then had a dismissive meeting with a hard-nosed editor at a leading university press who said bluntly: "There is nothing here." This made me understand why playwright/physician David Feldshuh in "Miss Evers' Boys" – a play and later an HBO movie on a fictional nurse meant to be Rivers – made up her moral dilemmas and imagined her speaking before the US Senate (which she never did).[18] There was just not enough in her own voice for a book. From another perspective, I started to worry that there was little I could add to what historian James H. Jones had already detailed in his masterful book on the Tuskegee study called *Bad Blood*, which Feldshuh had used to shape his play's main character.[19] As I was considering how to focus a book and what more to do with all the materials I had accumulated, the twentieth anniversary of the end of the study arrived and new documentary films appeared on PBS, ABC, and the BBC to mark the occasion.[20]

The documentaries included astounding interviews with physicians who had been involved in the study, and this would affect both my work and my activism. When asked about their thoughts on the study,

[17] S.M. Reverby, "Rethinking the Tuskegee syphilis study: Nurse Rivers, silence and the meaning of treatment," *Nursing History Review*, 7 (1999), 3–28; reprinted in S.M. Reverby (ed.), *Tuskegee's Truths*, pp. 365–385.

[18] D. Feldshuh, *Miss Evers' Boys* (New York, NY: Dramatists Play Service Inc., 1995). D. Feldshuh and W. Bernstein, *Miss Evers' Boys*, directed by Joseph Sargent (New York, NY: HBO Films, 1997).

[19] J.H. Jones, *Bad Blood: The Tuskegee Syphilis Experiment*, 2nd ed. (New York, NY: Free Press, 1993). J.H. Jones and D. Feldshuh had some legal dealings over what was borrowed from *Bad Blood*.

[20] Films for the Humanities and Sciences, *The Deadly Deception*, directed by D. Diianni (Boston, MA: WGBH Educational Foundation and NOVA – PBS, 1993). P. Craig, *Bad Blood*, directed by C. Gordon (London: A Diverse Production for Channel 4, in association with the Arts and Entertainment Network, 1992). J. Schadler (interviewer) and S. Olansky (interviewee), "The Tuskegee Study," *Primetime Live*, ABC News, February 6, 1992.

I imagined that it would have been simple for these physician-researchers to have answered, "It was a long time ago, it was the South, we know better now, so sorry, etc. etc." They did not.[21] It was as if they were still waving the science flag for their lost cause. In one particularly appalling scene in the PBS film, former US Public Health Service researcher Dr. John Cutler intoned: "My regret? In terms of the study I have none."[22] He and others who were interviewed seemed completely oblivious to what the study in Tuskegee had meant to African Americans and its relevance to bioethics. They continued to argue that there had been neither racism nor unnecessary research. Despite two decades of work by historians, bioethicists, lawyers, community members, and federal investigators, here were people in the research community who stood by their past and that of their colleagues to justify the research.

This lack of remorse and the failure of the government to issue an apology became the subtext of a 1994 University of Virginia symposium on the meaning of the study in Tuskegee, where one of the documentary films was shown.[23] As I have detailed elsewhere, many of us – scholars on the issues of race, history, anthropology, medicine, and experimentation – spoke at the symposium, and we were aghast at the physicians' statements in the film.[24] The University of Virginia's John Fletcher, one of the first bioethicists to have worked at the US National Institutes of Health, put away his lecture notes after he saw the film and talked about the need for a real apology, a way to make amends more publicly and to rebuke the physicians who still thought they had done right. We knew that fewer and fewer of the unknowing participants were still alive, and so there was urgency in considering what might be done.

Through various channels and with some money transferred from one project to another, public health/ethics researcher-dentists Ralph Katz and Rueben Warren convened a small group at Tuskegee University in

[21] For a similar contemporary view that argues that bioethics should "get out of the way" of science, see S. Pinker, "The moral imperative for bioethics," *The Boston Globe* (August 1, 2015). Available at: www.bostonglobe.com/opinion/2015/07/31/the-moral-imperative-for-bioethics/JmEkoyzlTAu9oQV76JrK9N/story.html. Accessed on: March 23, 2016.

[22] Films for the Humanities and Sciences, *The Deadly Deception.*

[23] "Doing bad in the name of good?: The Tuskegee Syphilis study and its legacy," University of Virginia Claude Moore Health Sciences Library Symposium (February 23, 1994). Available at: http://exhibits.hsl.virginia.edu/badblood/symposium/. Accessed on: October 11, 2016.

[24] S.M. Reverby, *Examining Tuskegee.*

January 1996, and an "Apology Committee" was formed.[25] A few months previous to this, in October 1995 President Bill Clinton had formally apologized to the thousands of victims of the US Cold War radiation experiments, and so it seemed reasonable to ask for an apology for the men who participated in the study in Tuskegee.[26] Our committee, now relabeled the Legacy Committee, wrote a short historical report explaining the need for an apology, and members of the Black Congressional Caucus and others began serious lobbying of the White House. Lobbying abated during the 1996 presidential campaign, but then continued in earnest after Clinton's reelection in November 1996. In April 1997, we learned that there would be an apology, although the event would be in Washington and not in Tuskegee, as we had asked.

The formal federal apology offered on May 17, 1997 (twenty-five years after public acknowledgment of the study) was a political spectacle, worthy of a tableau in a Southern novel.[27] I could not believe that our small "Apology Committee" had actually been part of making it happen and that I could be there for the occasion. Five of the six remaining men from the study in Tuskegee, including 103-year-old Fred Simmons, who had never flown before, walked in to thunderous applause in the formal and elegant East Room of the White House. The audience included the families of the research subjects, a political "Who's Who" of African Americans, "Legacy Committee" members, Tuskegee University leaders, and government officials. There was a direct video feed to Tuskegee for the hundreds who could not make the trip to Washington. Orchestrated with great dignity, study survivor Herman Shaw spoke first on behalf of the men. He made it clear that their tragedy should never happen again. Then, with his characteristic twinkle and no small sense of irony, he introduced the President.

President Bill Clinton was at his oratorical best as his Southern cadences were answered from the audience's amen corners. His words would have warmed any historian's heart. "Without remembering the past," he intoned, "we cannot make amends and we cannot go forward."

[25] J. Fletcher and V.N. Gamble, cochairs, "Legacy Committee request," in S.M. Reverby (ed.), *Tuskegee's Truths*, pp. 559–566.

[26] M. Cimons, "Clinton apologizes for radiation tests," *Los Angeles Times* (October 4, 1995). Available at: http://articles.latimes.com/1995-10-04/news/mn-53213_1_radiation-experiments. Accessed on: March 23, 2016. Advisory Committee on Human Radiation Experiments, *The Human Radiation Experiments* (New York, NY: Oxford University Press, 1996).

[27] S.M. Reverby, *Examining Tuskegee*, pp. 224–226.

And with great fervor he argued, "What the United States government did was shameful, and I am sorry." There was not a dry eye in the room. It was an emotional communication in full secular and religious terms of confession and repentance. The media coverage was enormous by the standards of the day (at a time when the social media we now have were not widely available).

What this apology did not do, however, was link "medical ethics and abuse of power" to racial injustice.[28] It was an apology for racism – something that was much needed – but not an apology for the ways in which scientific ideas are constructed regarding the assumptions about race.[29] This federal apology mattered, in the ways apologies do, because it acknowledged the pain and suffering of the men and their families and promised to make amends. For what and how, however, was left unspecified and, arguably, is still to be determined. It also gave no hint of how to make sure this never happened again, although the Legacy Committee asked for and got funding for a Bioethics Center at Tuskegee University.

The documentaries about the study in Tuskegee and the eventual federal apology brought home to me the need to consider the multiple ways in which stories of the study had been told, from conspiracy theories to scientific justifications.[30] I became increasingly interested in the ways in which historians were beginning to probe the questions of social memory and how cultures imagine their own pasts. Bringing the historian's concern with memory and reanalyzing the study and its "characters" seemed a worthy task given the study's renewed cultural currency.

I knew that a book to address these complicated issues would take more research and more time. Meanwhile, I had all this material that I had already collected. So as a first step I began work on a book project that would rely on much of primary and secondary material I already had in my files. I thought of this book as a teaching tool that would make the complexities of the study more visible and open to debate, and would allow readers to consider different points of view.[31] Why had the

[28] L.M. Harter, R.J. Stephens, and P.M. Japp, "President Clinton's apology for the Tuskegee syphilis experiment: A narrative of remembrance, redefinition, and reconciliation," *The Howard Journal of Communications*, 11 (2000), 19–34.

[29] W.J. Clinton, "Remarks by the President in apology for the study done in Tuskegee," in S.M. Reverby (ed.), *Tuskegee's Truths*, pp. 574–577.

[30] For examples of the differing views on the study, see S.M. Reverby (ed.), *Tuskegee's Truths*.

[31] S.M. Reverby (ed.), *Tuskegee's Truths*.

physicians thought that what they were doing was right? Were all the men really kept from being treated? What did we expect Nurse Rivers to do? The book was published as *Tuskegee's Truths: Rethinking the Tuskegee Syphilis Study*. My son thought that the book should have been called *Bad Blood 2: The Collection*, like a movie. I am fairly sure that historian James Jones would not have thought this particularly humorous. I hoped my presentation of the multiple viewpoints as primary documents would take readers past a kind of knee-jerk response and allow them to really consider what multiple individuals had thought and said. I wanted the study to live, not just be confined to the "bad old days" dustbin.

Out of concern for how this past had come to be understood, I struggled to shape what would become a second book, *Examining Tuskegee: The Infamous Syphilis Study and Its Legacy*, published years later in 2009.[32] In this book I retold the who/what/where/when/how of the study, and tried to explain why different subjects had been involved and what their research participation had meant in cultural and political terms. In the time since Jones had published *Bad Blood*, the men's medical records had become openly available in the federal archives and considerable statistical work was necessary to code and understand them. I worked really hard to do this work and I did many, many interviews with various groups to better understand their perspectives. The goal was to be able to explain the difference between what had actually happened and the particular ways in which the study was remembered.

My work was never just about my own research, however. I made contact with various groups memorializing the study in the Tuskegee area, gave historical advice, helped with fundraising, and spoke when requested. I gave lectures wherever asked, if time permitted, from the association of US Public Health Service officers to the families of men in the study. I reminded the Public Health Service physicians that their historical colleagues had been supported in their efforts and had thought that what they were doing was right. Needing a biblical reference when speaking at a small Baptist church just outside Tuskegee, I told the family members (from their church pulpit) that their relatives had wandered in a medical desert for forty years. I understood it to be my responsibility, as at Passover, to tell the story of what happened in Tuskegee over and over.

[32] S.M. Reverby, *Examining Tuskegee*.

I developed a website to support the arguments in the book and to share more of the statistical evidence.[33]

I knew, of course, that I could never control how others would think about or use the "Tuskegee" story. All I could do was provide the best history I could develop and share it with as wide an audience as possible. When the second book finally came out, I thought that my nearly twenty years of thinking about the history of syphilis were over. But this was not to be the case.

"Normal Exposure": The US Public Health Service Studies in Guatemala

While I was researching the book that would become *Examining Tuskegee*, I had gone to the University of Pittsburgh archives in April of 2003. I was on my way to visit family in West Virginia and thought I would make a quick stop. The archives contained the papers of Surgeon General Thomas Parran, the University of Pittsburgh's first dean of their public health school and the Public Health Service's leader most known for his work on syphilis in the 1930s. I was following the trail of another scholar's footnotes in an effort to understand some of the precursors to the study in Tuskegee.[34] I suspect that if these archives had been in Iowa or Idaho, I never would have made the trip since the information that I was tracking down was a bit peripheral to the arguments I was making in the Tuskegee book. I was just trying to be thorough. While I was at the archives, I was told that the papers of Dr. John Cutler, the former Public Health Service researcher who had defended the study in Tuskegee in the 1990s documentaries (and had taught at the University of Pittsburgh) were also archived there. Perhaps, I hoped, there would be more here to explain his views, or put them in context.

I found copies of published articles on the study in Tuskegee – and boxes and boxes of materials on a completely different study in Guatemala. In the first folder of the first box that I looked into, there was a typed report with the ominous title "Experimental studies on human inoculation with syphilis, gonorrhea, and chancroid" and the subtitle

[33] See S.M. Reverby, website for *Examining Tuskegee: The Infamous Syphilis Study and Its Legacy* (November 13, 2009). Available at: http://examiningtuskegee.com. Accessed on: May 31, 2016.

[34] H. Marks, *The Progress of Experiment* (New York, NY: Cambridge University Press, 2000).

"Experimental studies in prophylaxis against gonorrhea, syphilis, and chancroid following sexual exposure or inoculation."[35] These studies had taken place in Guatemala between 1946 and 1948, at a time when syphilologists were just coming to terms with what (the barely available) penicillin would mean for the treatment of what were then called venereal diseases. This was before Cutler was involved in the study in Tuskegee, and while he was in the Public Health Service.

To this point in my career, I had spent nearly a dozen years explaining that no one in the study in Tuskegee had been *given* syphilis by the government. The men were already infected. But here was a story about Cutler, the Public Health Service, syphilis, inoculation in a Global South country intentionally exposing subjects to sexually transmitted diseases (STDs), and support for these studies coming from the National Institutes of Health, the Pan American Sanitary Bureau, and the Guatemalan Ministry of Health.[36] Then again, the subtitle for this research made reference to "prophylaxis." This suggested that there had been some kind of intervention to prevent disease, unlike in the study in Tuskegee where the Public Health Service had tried (not always successfully) to prevent the participants from receiving treatment. I could not stop reading.

The research had been done in a central penitentiary in Guatemala City, in an army barrack, and in a mental hospital and was complicated by the necessity of figuring out how to transmit syphilis and other STDs to the subjects.[37] It started with sex workers and prisoners. Prostitution was legal in Guatemala at the time, and it was also possible to bring sex workers into the prison to ply their trade. The research involved hiring infected street sex workers as the "vectors" to transmit syphilis, and following up with blood tests to see if disease had been transmitted to the prisoners. In addition, Cutler and his colleagues added the provision of alcohol and cigarettes as incentives to the prisoners in what he deemed "normal exposure."[38] There had been studies on occasion in

[35] Box 1, John C. Cutler Papers (Pittsburgh, PA: University of Pittsburgh Archives). Cutler's papers were moved in 2010 to the Southeast Regional National Archives in Morrow, Georgia.

[36] Presidential Commission for the Study of Bioethical Issues, *"Ethically Impossible" STD Research in Guatemala from 1946 to 1948* (Washington, DC: US Department of Health & Human Services, September, 2011). Available at: http://bioethics.gov/node/654. Accessed on: May 31, 2016.

[37] "Untitled Report" (February 24, 1954), Box 1, Folder 1, John C. Cutler Papers.

[38] "John C. Cutler to John F. Mahoney" (September 18, 1947), Box 1, Folder 13, John C. Cutler Papers.

the United States that had used sex workers in this way, but mostly this kind of research did not occur in the United States.[39]

The study in the Guatemalan prison then morphed into a study of mental patients in the country's only asylum, and soldiers in an army barrack. As well, there was additional research in the prisons involving inoculation through means other than sexual intercourse, and there was also some blood testing, but not inoculations, at an orphanage. I could tell from the correspondence, photographs, and reports that the researchers had been using inoculations in some horrific ways: scarification of penises and cervixes, arms, faces, spinal injections. I could hardly believe what I was seeing and reading; it was all overwhelming. There was no real research protocol (as there had not been in Tuskegee nor in many studies at the time), and except for a few serology reports, nothing about the study in Guatemala ever explicitly made it into published research journals.[40] Information about this study had not shown up in all the research I had done on syphilis studies in this time period, nor did I know about it from the historical or bioethics literature.

At first glance, however, it did appear that the researchers in Guatemala were treating infected persons. One summary of treatment results noted that "all were given penicillin"; another stated that "very active homosexual ... represents an instance of reinfection and so as not to have a potentially infected individual he was retreated with a dose of penicillin identical with that originally administered."[41] Cutler was writing to his Public Health Service supervisors: "For purposes of facilitating our work here, Dr. Spoto [Pan-American Sanitary Bureau

[39] "Thomas Parran to Joseph Earle Moore" (March 24, 1938), Box 5, Folder 23, Thomas Parran Papers (Pittsburgh, PA: University of Pittsburgh Archives). "Joseph Earle Moore to Thomas Parran" (April 4, 1938), Box 5, Folder 23, Thomas Parran Papers.

[40] S. Levitan, H.A. Aragon, J.C. Cutler, J.M. Funes, J. Portinoy, and A. Paredes-Luna, "Clinical and serologic studies with reference to syphilis in Guatemala Central America. I. Studies of comparative performance of the Kahn, Kolmer, Mazzini, and VDRL slide tests as carried out in the national orphanage," *American Journal of Syphilis, Gonorrhea, and Venereal Diseases*, 36:4 (July 1952): 379–387. J.C. Cutler, S. Levitan, R.C. Arnold, and J. Portinoy, "Studies on the comparative behavior of various serologic tests for syphilis. II. A report on an observed pattern of entrance into seroreactivity among patients with untreated primary syphilis," *American Journal of Syphilis, Gonorrhea, and Venereal Diseases*, 36:6 (November 1952), 533–544. J. Portinoy, R. Galvez and J.C. Cutler, "Clinical and serologic studies with reference to syphilis in Guatemala, Central America. III. Studies of comparative performance of the Kahn, Kolmer, Mazzini, and VDRL slide tests among leprosy patients," *American Journal of Syphilis, Gonorrhea, and Venereal Diseases*, 36:6 (November 1952): 566–570.

[41] "Summary of results," Box 1, Folder 4, John C. Cutler Papers.

Caribbean sector chief] believes it very necessary for us to undertake some treatment programs."[42] A few weeks later Cutler would report, "Both Dr. Spoto and I are anxious to get our testing and treatment program well under way before we start prophylactic work."[43] Given that there appeared to be some kind of treatment, even if there was clearly exposure/infection without consent, it seemed different from what had happened in Tuskegee.

At the time, I did not have a clear idea about what to make of the information that I had uncovered. Without a huge statistical analysis, I could not make sense of the myriad reports on the individual subjects to determine what really had happened. How many subjects were treated? How many were infected? I had no slew of graduate students, no training in biostatistics, and a medical historian's understanding of syphilis. I kept my notes and photocopies, and I spoke to a few people about this research. I was well aware there were many horror stories about medical research, and in all honesty, I did not think through why this story might be different.

Things changed after I interviewed David Sencer. The longest running director of the Centers for Disease Control and Prevention (CDC; 1966–1977), Sencer had convened a meeting in 1969, after criticisms of the study in Tuskegee mounted internally, to discuss whether that study should be continued. He sent two public health advisors to Tuskegee to speak to the nearly all African American local medical society.[44] In the end, he decided there was still enough support to continue the study in Tuskegee, and he could find no other way to provide the small amounts of money that were still being given to some of the families. The irony is that if he had ended the study, then we probably would not have known about it since there might not have been any media outcry. Much of what was known on the study happened because of the media outcry, Fred Gray's lawsuit, and historian Jim Jones's calling Gray to say he knew the location of the papers in the National Archives of the study in the 1930s. Jones, however, had assumed the study had ended in the

42 "John C. Cutler to John F. Mahoney" (September 3, 1946), Box 1, Folder 13, John C. Cutler Papers.

43 "John C. Cutler to John F. Mahoney" (November 12, 1946), Box 1, Folder 13, John C. Cutler Papers.

44 There are no minutes of that meeting. I interviewed one of the advisors who said there was agreement to continue the study in Tuskegee, but a physician who was there was much more vague about what really went on. See S.M. Reverby, *Examining Tuskegee*, pp. 217–219.

1930s. Without the outcry and the lawsuit, Jones would never have known it had continued nor made contact with Gray. Without the connections to Gray, much of Jones's story would have been more limited.[45] When the story of Tuskegee was made public in 1972, protestors burned Sencer's effigy and "Tuskegee" began its move into the American consciousness.[46]

When I went to Atlanta in June 2003 to code the Tuskegee medical records in the archives, Sencer and I had dinner together. He was charming and thoughtful, although he clearly knew that what he had done about Tuskegee was wrong. In his mid-seventies by then, he really cared about history and what the CDC had done; indeed, the CDC museum was named after him after he passed away in 2011. He even seemed to understand that had I been in Atlanta in 1972, I would have been outside his office in protest. After that initial dinner, we continued to write back and forth via email, and to see one another on occasion when I came through Atlanta. One time, in person, he said to me, "I cannot believe anyone thinks we would have given the men syphilis in Tuskegee." I quickly responded, "Well, Dave, there is Guatemala." He looked a bit shocked and said, "What do you know about Guatemala?" And I shot back, "What do you know about Guatemala?" He explained that whatever happened, it was before he was even out of medical school, but he admitted that there had been rumors about Guatemala in the sexually transmitted diseases division of the CDC. As far as he knew, there were no records at the CDC. I had been through the CDC records of the Sexually Transmitted Diseases division at the National Archives in Morrow, Georgia, and knew there was nothing there, and no publications. I told him I had found information about what happened in Guatemala in the Cutler papers and that eventually I would be writing this up.

When the *Examining Tuskegee* book was done in 2008, more than one hundred pages were edited out including my preliminary information on the Guatemala material that I realized would need an article of its own. In June 2009, I returned to the University of Pittsburgh archives to redo the research (to confirm my findings) for what I thought would be a paper on the study in Guatemala as my final work on this subject. I had no intention of trying to code and make sense of the thousands of lab and research reports on the actual subjects' experiences. In the end, because

[45] S.M. Reverby, interview with Jim Jones and Fred Gray, Montgomery, AL, May 1998.
[46] S.M. Reverby, *Examining Tuskegee*, pp. 81–82, 207–208, 291–292.

of personal and professional commitments, it was not until March 2010 that I finished drafting what would become a paper presented at the American Association for the History of Medicine Annual Meeting in May 2010, and later an article in the *Journal of Policy History*.[47] My text linked the syphilis studies in Tuskegee and Guatemala, explained the differences, and tried to point out the structural and scientific reasons why the study in Guatemala had happened. I could not, and did not, figure out how many people *really* had been infected, or how many had been treated.[48]

As most historians of medicine who are not physicians know, if there is one medical or scientific error in a publication, it will be criticized (and possibly even dismissed) even if the analysis is correct. To avoid any such mistakes, in March 2010 I asked Sencer (then in his eighties), as my syphilologist colleague/friend, to review the science in the paper I had drafted. After reading my draft he called me (we usually communicated by email) and said, "What we have here is a political problem, not a scientific problem." He was deeply worried that once news of the paper came out, his beloved CDC would be covered in infamy again. Having judged the Tuskegee story incorrectly before, he had no intention of making a similar mistake. I naively said, "But Dave, it is going to be in an obscure history journal." His reply was prescient, "No, in the age of the internet, once this is out there, everyone will be concerned." He asked if he could share my draft with the CDC higher-ups. I did not frankly think of taking any of this to the media myself at the time, and was willing to see what would happen within the government bureaucracy first. A lone historian saying this happened versus a governmental response of some sort seemed worth waiting for, at least for a bit of time. Historians find this kind of horror story in medical research often, unfortunately, and I had no way of knowing if this would be any different from any others.

I had no reason to stop the CDC from looking at my materials. I do not work for the CDC, nor do I ever plan to. They have no control over what I write or say, and the paper had already been accepted in May 2010 for publication in January 2011. During the early summer of 2010 there was a furor of activity. I spent a good deal of time on the phone with several CDC officials (primarily then-division chief of the Sexually Transmitted Diseases section of the CDC John Douglas,

[47] S.M. Reverby, "'Normal exposure' and inoculation syphilis: A PHS 'Tuskegee' doctor in Guatemala, 1946–1948," *Journal of Policy History*, 23:1 (January 2011), 6–28.

[48] S.M. Reverby, "Normal exposure."

Associate Director for Science Harold Jaffe, as well as their crisis management person) and Sencer. The discussions focused on what should happen and the need for a federal response before my article was published. I sent all of my notes (after correcting spelling errors) and photocopies to Sencer's house, not the CDC. This was a precaution in case nothing ever happened and someone made a Freedom of Information Act request to the CDC. I had the sense that the CDC wanted to do something about this before my paper was published, but I was not sure what would happen. My notes were clearly not enough. Next, Jaffe sent Douglas to the Pittsburgh University archives to look at the materials. Douglas brought back much of the data for his staff to analyze, and he wrote his own report.[49]

Shortly thereafter, negotiations took place within the CDC hierarchy (including with their Guatemala-based regional director for Central America), then with groups in the National Institutes of Health, and finally with the White House Domestic Policy Council and the State Department. One of these days I will do a Freedom of Information Act request to find out exactly what happened, some of which Sencer could tell me, some of which he likely could not. Key to the negotiations concerning what would or would not happen was Ezekiel Emanuel, the physician and bioethicist heading President Obama's health initiatives (and brother of Chicago mayor Rahm Emanuel, then President Obama's chief of staff). Ezekiel Emanuel had a long history in bioethics, knew my scholarship, and had recently coedited a book entitled *Exploitation and Developing Countries: The Ethics of Clinical Research*.[50] He was working with others in the administration on changing research ethics policy.

What they would all do about my article and Douglas's report was unclear. At first, it looked like there might just be a meeting at the CDC to discuss what we had both found. Somewhere along the line, however, it must have become obvious to someone in the governmental health bureaucracy and then the State Department that it was important to

49 Centers for Disease Control and Prevention, *CDC Report on Findings from the U.S. Public Health Service Sexually Transmitted Disease Inoculation Study of 1946–1948, Based on Review of Archived Papers of John Cutler, MD, at the University of Pittsburgh* (September 29, 2010). Available through US Department of Health & Human Services web page, "Information on the 1946–1948 United States Public Health Service STD inoculation study." Available at: www.hhs.gov/1946inoculationstudy/std_inoc_study.pdf. Accessed on: October 19, 2016.

50 J.S. Hawkins and E.J. Emanuel (eds.), *Exploitation and Developing Countries: The Ethics of Clinical Research* (Princeton, NJ: Princeton University Press, 2008).

extend a real apology to the people of Guatemala and that this should be done *before* my article was published that coming January 2011.

Several steps were taken to make this happen. The White House orchestrated the federal response. President Obama would call Alvaro Colom, the President of Guatemala, to explain what had happened, and Secretary of State Hillary Rodham Clinton and Secretary of Health and Human Services Kathleen Sebelius would issue a formal apology. The Presidential Commission for the Study of Bioethical Issues would be asked to investigate the matter and issue a report.[51] There would be a brief press conference and materials would be prepared for the web. I would post my article (which had been accepted for publication, but not yet copy-edited) on my faculty webpage at Wellesley College, where I taught. I would do this at 9:00 a.m. on October 1, 2010. The health staffers in the House of Representatives and the Senate were alerted to the story. This last piece of information I knew because my daughter, who worked at the time for Representative Steny Hoyer, called me the night before the public announcement to say this would be a really big story. I did not quite believe her.

I was at home the morning of October 1, 2010. There was no plan for me to be at a press conference, but my cell phone number was listed on my college webpage as the press contact just in case. I thought, along with the college's public relations person, that we would have the weekend for the story to gain what journalists call "legs." I put a copy of my article up on my faculty page at 9:00 a.m. At 9:01 a.m., a media story went out from Robert Bazell, the NBC science correspondent, on MSNBC's website.[52] He had been given the embargoed story the night before, a fact the government officials had not told me. If I had known this, I would have had more time to prepare a response and certainly not given out my cell number. At 9:05 a.m. Bazell phoned me, and by 9:30 a.m. there was an NBC news crew at my house. At 11:00 there was a press conference with National Institutes of Health and State Department bigwigs for the media. Then there was very little coming from the government for the next few hours.

The rest of the world's media seemed to be calling my cell phone all that day and night while I parried emails. Most of the media attention

[51] Presidential Commission for the Study of Bioethical Issues, *Ethically Impossible.*

[52] R. Bazell, "U.S. Apologizes for Guatemala STD Experiments," *MSNBC* (October 1, 2010). Available at: www.nbcnews.com/id/39456324/ns/health-sexual_health/t/us-apologizes-guatemala-std-experiments/#.WAPqYlvt_O8. Accessed on: October 16, 2016.

focused on the use of American taxpayer money to support sex workers and prisoners having sex, the purposeful transmission of sexually transmitted diseases, the interactions of our government with a Global South country whose government it would overthrow in 1954, the awful descriptions of what amounted to medical torture, and the connection between the study in Guatemala and the study in Tuskegee. It was also big news that the US government at a high level was apologizing to another country. The media attention continued for a few weeks. I was peppered with requests for speaking engagements and was consulted (briefly) by the Presidential Commission (I went to several of their meetings). Over the next five years, I think I gave over fifty different lectures (in the United States, Guatemala, and England) dealing with this topic. I wrote several short articles, did hundreds of interviews, and tried to keep the issues alive. I had a sense there was limited time to take advantage of the media attention and that I had to do this now.

Mostly I tried to talk about why seeing the studies in Guatemala as merely a melodrama of an infamous and cruel American doctor in a Global South country would not explain what happened.[53] I attempted to explore why Cutler and his superiors thought this was appropriate, if difficult, research to do, and why they kept it quiet.[54] I tried to remind medical personnel how easily they *could* be Cutler rather than imagining they *never could* be him. I did more research on Cutler, and met privately with his family members who contacted me. I felt responsible for how he would go down in history, and I wanted to get it right so that the story did not become one of simple good and evil.[55] I primarily focused on the problems of exploitation, why Guatemalan health officials would have said yes to the study, and why the Public Health Service in the United States thought it was important to do this work. The Presidential Commission's staff spent a year on their research, filled in many of the details, provided a wider analysis, was able to access materials I had not found, and went to Guatemala. Their two reports, *"Ethically Impossible" STD*

[53] S.M. Reverby, "Ethical failures and history lessons: The U.S. Public Health Service Research studies in Tuskegee and Guatemala," *Public Health Reviews*, 34:1 (2013), 1–18. S.M. Reverby, "Enemy of the people/enemy of the state: Two great(ly infamous) doctors, passions, and the judgment of history," *Bulletin of the History of Medicine*, 88:3 (Fall 2014), 403–440.

[54] L. Reynolds, "Susan Reverby: I can't let Cutler become a barbarian," *Plaza Pública* (January 26, 2015). Available at: www.plazapublica.com.gt/content/i-cant-let-cutler-become-barbarian. Accessed on: March 23, 2016.

[55] S.M. Reverby, "Enemy of the people."

Research in Guatemala from 1946 to 1948 and *Moral Science: Protecting Participants in Human Subjects Research*, were made public with links to the primary materials and a study guide.[56] Others have since explored more fully how structural factors about race, gender, and sexuality affected the research.[57] As well, more nuanced articles on prophylactic research and the scientific/power assumptions have appeared.[58] Graduate students and a few other researchers have been in touch with me about this research, and I expect there will be more articles, maybe even dissertations, on this topic in the years to come.

Today, six years later, it is not clear to me what, if any, impact this knowledge has had outside a narrow swath of the academy in the United States; in Guatemala it has become another on a long list of outrages. Slowly, however, the story seems to be making it onto lists of the horrible things the American government has done in the name of science (that lives on the Internet, if nowhere else).[59] Unlike the study in Tuskegee where there is a local community that lives the memory of this study and passes it into the African American population, there is no large Guatemalan community in the United States to carry on the memory of that study. There have been a few short filmed news stories, but it is not unusual for some of the facts to get really muddled.[60] The Presidential Commission did prepare a study guide to assist in explaining what happened.[61] The difficulty in getting all the facts right is clear in one of the more commonly used vignettes. While there is no evidence that any of the children in the orphanage, who were examined and treated by

56 Presidential Commission for the Study of Bioethical Issues, *Ethically Impossible.*

57 C. Galarneau, "'Ever vigilant' in 'ethically impossible': Structural injustice and responsibility in PHS research in Guatemala," *Hastings Center Report*, 43:3 (May–June 2013), 36–45. H. Fernandez-Lynch, "Ethical evasion or happenstance and hubris? The U.S. Public Health Service STD inoculation study," *Hastings Center Report*, 42:2 (March–April 2012), 30–38.

58 I. Lowy, "The best possible intentions: Testing prophylactic approaches on humans in developing countries," *American Journal of Public Health*, 103:2 (February 2013), 226–237.

59 "The 13 Most Evil U.S. Government Human Experiments," *TopInfo Post* (n.d.). Available at: www.topinfopost.com/2014/10/12/the-13-most-evil-u-s-government-human-experiments. Accessed on: October 15, 2016.

60 See, for example, SBS, "Infected!," *Australia SBS Dateline* (June 19, 2011). Available at: www.sbs.com.au/dateline/story/about/id/601221/n/Infected. Accessed on: March 23, 2016.

61 Presidential Commission for the Study of Bioethical Issues, "A Study Guide to 'Ethically Impossible' STD Research in Guatemala from 1946 to 1948." Available at: http://bioethics.gov/node/778. Accessed on: October 15, 2016.

American physicians, were infected with a sexually transmitted disease, that story continues. The photograph of a woman who was in the orphanage and remembers having a gringo doctor do a vaginal exam (which is possible) is the face of the research. But whether she or others in the orphanage were infected is not clear in the extant paper records.[62] The government report also made clear that in the end they could not determine accurately how many people had really become infected or treated.[63]

An initial lawsuit in the United States found a number of people in Guatemala who claimed they had been harmed by the studies sponsored by the government of the United States and tried to sue the government under the Alien Tort Statute of 1789, which allows foreign nationals to sue for human rights violations. The case was dismissed because the judge claimed that however "troubling [the facts of the case] ... federal law bars claims against the United States based on injuries suffered in a foreign country."[64] In the spring of 2015, another group of lawyers found 750 individuals to act as plaintiffs, including people presumed to be children, spouses, and grandchildren of people in the studies in Guatemala. Instead of suing the US Public Health Service, they sued, again in the United States, Johns Hopkins University and the Rockefeller Foundation, whose researchers approved, monitored, and were involved in the studies in Guatemala. They also sued the pharmaceutical giant Bristol-Meyers-Squibb, whose predecessor company supplied the penicillin used in the studies.[65] They are asking for a staggering one billion dollars in compensation. This case is still wending its way through the court system and is proceeding.[66] Finally, in December 2015, the Archdiocese of

[62] See, for example, R. Ellis, "Guatemalans deliberately infected with STDs sue Johns Hopkins," *CNN* (April 14, 2015). Available at: www.cnn.com/2015/04/03/americas/guatemala-std-lawsuit/. Accessed on: March 23, 2016.

[63] Presidential Commission for the Study of Bioethical Issues, *Ethically Impossible*, p. 154.

[64] Associated Press, "Lawsuit over U.S. Guatemala syphilis experiment dismissed," *CBCNews/World* (June 13, 2012). Available at: www.cbc.ca/news/world/lawsuit-over-u-s-guatemala-syphilis-experiment-dismissed-1.1216241. Accessed on: March 23, 2016.

[65] O. Laughland, "Guatemalans deliberately infected with STDs sue Johns Hopkins University for $1bn," *The Guardian* (April 2, 2015). Available at: www.theguardian.com/us-news/2015/apr/02/johns-hopkins-lawsuit-deliberate-std-infections-guatemala. Accessed on: March 23, 2016.

[66] In September 2016, a federal judge in Baltimore dismissed the suit, but the lawyer for the defendants is pledging to continue it; see S. Gantz, "Judge dismisses $1 billion Guatemalan syphilis experiment case against Hopkins, others," *Baltimore Sun* (September 9, 2016). Available at: www.baltimoresun.com/business/bs-bz-guatemala-lawsuit-dismissed-20160909-story.html. Accessed on: October 15, 2016.

Guatemala sued the United States and Guatemala at the Inter-American Commission on Human Rights for "human rights violations and crimes against humanity."[67] Whether any of this will bring any monetary compensation to the studies' victims and their families remains unclear.

"So What?" and Social Change

What in the end, then, might we learn here? In my *Examining Tuskegee* book I argued that we needed to separate out what happened from what is remembered to have happened. I hoped that, with the facts and complexities at least somewhat knowable, the use of "Tuskegee" as a metaphor for medical racism would no longer be needed. I have similar hopes about the studies in Guatemala.

I know that in each of these cases the apologies mattered, and that doing careful history was crucial to making these happen. It matters that we get the facts and complexities as correct as possible and that we not reach for easy explanations in what I keep calling the "melodrama" of medical experimentation that just focuses on evil men and their victims. The righting of injustice in these cases required at least beginning with a serious acknowledgment of what really happened. I also know, however, that we cannot count on academic work alone to make a difference. For example, in these cases, just producing academic work probably would not have gotten media attention, and without media attention it very likely would not have been possible to get the attention of the general public and the government (which is what eventually led to the apologies). But the question about "so what?" continues to haunt me.

I am also aware of the historical contingencies that made all of this happen. If the films had not been made on the twentieth anniversary of the study in Tuskegee and if many of us had been unable to see John Cutler saying he was not sorry, the push for the apology for that study might not have happened. If I had not followed another scholar's notes back to the archives in Pittsburgh, and then been told that Cutler had materials there, I have no idea how long it might have taken for someone

[67] R. Garcia, H. Robinson, and C. Sweetser, "Archdiocese of Guatemala files international petition against US and Guatemala for human rights violations and crimes against humanity STD experiments," *The City Project* (December 14, 2015). Available at: www.cityprojectca.org/blog/archives/41463. Accessed on: May 31, 2016.

else to find and use his papers on what happened in Guatemala. If I had not made friends with David Sencer, and if he had not had his particular history with Tuskegee and the CDC, my article might simply have appeared in an academic journal with no effect.

I have been criticized for not telling the Guatemala story sooner, for example, when I first learned about it in 2003.[68] Perhaps I should have done so then, but would I have been able to solicit any kind of meaningful response at this time? Would then-President Bush have done anything in response to the concerns of a lone academic? I think it mattered that when I did raise concerns about the studies in Guatemala, my book *Examining Tuskegee* had been published, but maybe not. It clearly did matter that Sencer (with his powerful connection to the CDC) was involved, and in 2003 I had not yet gotten to know him well enough to be in close contact.

We are always faced with what can be done to prevent harms, and what kinds of redress there should be "for historical injustices in the United States."[69] Saying "I'm sorry" matters, but only so much, for it tells us little about what should happen next. As literary critic Gayatri Chakravorty Spivak has argued, "injustice" needs "righting wrongs," not just "having or claiming a right or a set of rights."[70] For many, this has meant a demand for reparations, through lawsuits and legislation, to fund care for those who have been harmed in medical practices and research. Such forms of justice do not always prevent other injustices from happening, but they do provide a form of restorative history and at least some examples that such kinds of work will be costly for institutions and individuals who did it.[71]

If there is to be a good answer to the "so what" question, then doing the best history we can matters, but this will never be enough. For there to be social change and justice, careful history has to be followed by political alliances, media coverage, and a strategy to keep the issues front

68 S.E. Lederer, "Shining light on a shady study," *Hastings Center Report*, 42:2 (March–April 2012), 3.

69 M.T. Martin and M. Yaquinto (eds.), *Redress for Historical Injustices in the United States* (Durham, NC: Duke University Press, 2007).

70 G.C. Spivak, "Righting Wrongs," *South Atlantic Quarterly*, 103:2–3 (2004), 523–581, at p. 523, quoted in M.T. Martin and M. Yaquinto, "On redress for racial injustice," in M.T. Martin and M. Yaquinto (eds.), *Redress for Historical Injustices*, pp. 1–32, at p.2.

71 S.M. Reverby, "Restorative justice and restorative history for the sexually transmitted disease inoculation experiments in Guatemala," *American Journal of Public Health*, 106:7 (July 2016): 1163–1164.

and center for a very long time. I share political theorist Mahmood Mamdani's view that "atrocity cannot be its own explanation. Violence cannot be allowed to speak for itself, for violence is not its own meaning. To be made thinkable, it needs to be *historicized*."[72] My worry is about who will carry on explaining these studies in new ways, and thereby continue to make them relevant to contemporary concerns.

[72] M. Mamdani, *When Victors Become Killers* (Princeton, NJ: Princeton University Press, 2001), quoted in V. Sanford, *Buried Secrets: Truth and Human Rights in Guatemala* (New York: Palgrave Macmillan, 2004), p. 14.

4

Twenty Years of Working toward Intersex Rights

ALICE DREGER

Like a lot of academics who have become advocates for patients and medical research subjects, I took up intersex rights work two decades ago because what was occurring in clinics looked so wrong – so patently irrational and unethical – it not only seemed *necessary* to change, it seemed it would be *easy* to change. Unexpectedly, it hasn't been easy to change. Today, in response to the intersex rights movement, human rights organizations all over the world, right up to various programs of the United Nations, have condemned what happens to intersex children in pediatrics hospitals worldwide.[1] Yet the condemned practices continue as "standard of care."[2] This article attempts to parse out why that is and mostly to describe some of how I've helped, as a historian of medicine and sexuality, with the intersex rights movement.[3]

First, some basic background. "Intersex" is a blanket term sometimes used for a wide variety of developmental sex anomalies. There are at least three dozen conditions we can call intersex; by definition, they all involve being born with a body that medical professionals have decided does not count as typically male or typically female. "Intersex" can refer to

[1] For example: United Nations Human Rights Office of the High Commissioner, "Fact sheet: Intersex," Free & Equal – United Nations for LGBT Equality (2016). Available at: https://unfe.org/system/unfe-65-Intersex_Factsheet_ENGLISH.pdf. Accessed on: May 24, 2016. J.E. Méndez, "Report of the Special Rapporteur on torture and other cruel, inhuman or degrading treatment or punishment," United Nations General Assembly, Human Rights Council, 22nd session, Agenda item 3, A/HRC/22/53 (February 1, 2013). Available at: www.ohchr.org/Documents/HRBodies/HRCouncil/RegularSession/Session22/A.HRC.22.53_English.pdf. Accessed on: May 24, 2016.

[2] For example: P.D.E. Mouriquand, D.B. Gorduza, C.-L. Gay, H.F.L. Meyer-Bahlburg, L. Baker, L.S. Baskin, et al., "Surgery in disorders of sex development (DSD) with a gender issue: If (why), when, and how?," *Journal of Pediatric Urology*, 12:3 (2016),139–149.

[3] Alice Dreger, "Cultural history and social activism: Scholarship, identities, and the intersex rights movement," in F. Huisman and J.H. Warner (eds.), *Locating Medical History: The Stories and Their Meanings* (Baltimore, MD: Johns Hopkins University Press, 2004), pp. 390–409.

ambiguous genitalia (genitals that develop in utero to be between the male and female types), but can also refer to when a person is born with one apparent sex externally but parts of the other sex internally. It sometimes involves unusual "sex" chromosome combinations, like having XXY (Klinefelter syndrome) or being a genetic mosaic (for example, having XX in some of your cells and XY in others). But sex development is far more complicated than "sex chromosomes"; some genetic females are born quite male in terms of anatomy and physiology, some genetic males are born quite female, and some are born with a lot of sex-type blending. Most people with intersex have either XX (typical female) or XY (typical male) chromosomes, and in many cases, the root cause of intersex anatomy development is never found. About 1 in 1,500 babies is born with notably atypical genitals, and about 1 in 200 babies is born with genitals different enough from typical that a surgeon may be called for a consultation. But some forms of intersex are not detected until puberty, adulthood, middle age, or even old age.

"Intersex" is a term that is over a hundred years old in science and it is used to talk about biological reality, but today it is also specifically a *political* term and category, as I explain below. The term was selected by activists in the early 1990s when they decided to form a movement that would be in part about rejecting the more common and much more stigmatizing medical parlance in their medical records: "hermaphrodite" and "pseudo-hermaphrodite."[4]

In the early 1990s, medical professionals were using a system of care that had been developed under the founder of pediatric endocrinology, Lawson Wilkins, at Johns Hopkins University, starting in the 1930s. Wilkins had recognized that the culture demanded that everyone fit into a male or female category. In the nineteenth century, physicians had tried to discern the "true sex" of apparently hermaphroditic patients,[5] but Wilkins and other physicians of his time took a clinically pragmatic approach to the problem of visibly intersex people, assigning gender based on what worked rather than on what the "true sex" appeared to be. When dealing with intersex children, Wilkins tried to figure out which gender they were likely to pass as with medical assistance, assigned them that gender if their social gender identity had not already been

[4] C. Chase, "Hermaphrodites with attitude: Mapping the emergence of intersex political activism," *GLQ: A Journal of Gay and Lesbian Studies*, 4:2 (1998), 189–211.

[5] A.D. Dreger, "Hermaphrodites in love: The truth of the gonads," in V.A. Rosario (ed.), *Science and Homosexualities* (New York: Routledge), pp. 46–66.

irrevocably established, and then arranged what medical care he could to help them pass. (In cases where intersex children had life-threatening hormonal conditions, Wilkins also did pioneering endocrine research and treatment that ultimately saved thousands of lives.)[6]

In the 1950s, Wilkins's approach was given a theoretical developmental framework by three psychologists who had joined his practice: John Hampson, Joan Hampson, and John Money. Wilkins's approach came to be dubbed the "optimum gender of rearing" method, and Money in particular pushed that method very hard, claiming to provide clinical proof from many case studies that gender identity and sexual orientation were primarily products of upbringing (nurture, not nature). Curiously, Money's own dissertation research at Harvard University on intersex adults who had lived without "corrective" genital surgeries in childhood had actually showed that most individuals born with intersex did well psychologically, adjusting as typical men and women without surgeries, with lower rates of psychopathology than the general population.[7] But Money was committed to being seen as a major theoretician in psychology, and so he worked to use the intersex clinical population to prove his theory that gender development and sexual orientation are matters of nurture, not nature. He did this by supporting radical genital surgeries designed to make intersex children's genitals look more typically male or female. The idea was that children who were visibly intersex would not develop "correct" gender identities (as straight men and straight women), whereas children who were visibly of one sex could be nurtured into the corresponding gender identities.

Along with using cases of many intersex children to prove his theory of gender and sexuality development, Money famously used a case of surgical sex change in early childhood in a non-intersex boy to "prove" that gender and sexual orientation were produced by nurture, not nature. At Money's recommendation, a non-intersex identical twin baby boy was surgically changed to a female following the accidental destruction of the boy's penis. Money claimed that the child grew to feel like a girl and to be a socially convincing girl, with the interests and behaviors of typical girls.

[6] S. Eder, "From 'following the push of nature' to 'restoring one's proper sex': Cortisone and sex at Johns Hopkins's Pediatric Endocrinology Clinic," *Endeavour*, 36:2 (June 2012): 69–76.

[7] J. Colapinto, *As Nature Made Him: The Boy Who Was Raised as a Girl* (New York: Harper Collins, 2000).

Thanks to the work of sex researcher Milton Diamond[8] and journalist John Colapinto,[9] however, we now know that Money misrepresented the case. Their reports on the true life history of the sex-changed twin, ultimately named David Reimer, showed that he did not adjust as a girl and ultimately identified as a straight man.

While Money's defenders have often noted that his papers supported eventually telling surgically altered children what had happened to them, in practice, clinicians avoided disclosing medical histories to intersex patients out of fear of upsetting them (and also very likely because the clinicians had not processed their own confusion and shame around sex). In 1995, an undergraduate medical student, Anita Natarajan, argued it was more ethical to lie to women born with androgen insensitivity syndrome (XY chromosomes, testes, and physiological insensitivity to "masculinizing" hormones) about their histories than to tell them the truth, because the truth would supposedly be too risky.[10] The essay was the second-place winner in the annual *Logie Medical Ethics Essay* Contest (sponsored by the *Canadian Medical Association Journal*) and with this award Natarajan received a $1,000 cash prize. (In 2001, she issued a clarification and apology.[11])

Patients' rights movements are built around shared diagnoses – e.g., HIV, breast cancer. What made the intersex rights movement unusual was that most of the people implicated by it had not been told their real diagnoses and medical histories, which meant they did not even know they were implicated. "Sparing" intersex patients the whole truth was, in practice, pretty much the norm until at least the late 1990s. When I started interviewing intersex people in the mid-1990s, many told me, "I thought I was the only one." Some were suffering unnecessarily from lack of medical care, because they were so traumatized by what had happened to them – particularly by repeated genital displays and genital

[8] M. Diamond and H.K. Sigmundson, "Sex reassignment at birth: Long-term review and clinical implications," *Archives of Pediatrics & Adolescent Medicine*, 151:3 (March 1997), 298–304.

[9] J. Colapinto, *As Nature Made Him*.

[10] A. Natarajan, "Medical ethics and truth-telling in the case of androgen insensitivity syndrome," *Canadian Medical Association Journal*, 154:4 (1996), 568–570. S.A. Groveman, "Sex, lies and androgen insensitivity syndrome," *Canadian Medical Association Journal*, 154:12 (June 1996), 1829 and 1832.

[11] A. Natarajan, "Medical ethics and truth-telling in the case of androgen insensitivity syndrome" Letter from the author, five years later. Availabe at: https://www.collectionscanada.gc.ca/eppp-archive/100/201/300/cdn_medical_association/cmaj/vol-154/0568e.htm. Accessed on: February 19, 2018.

exams for trainee groups in teaching hospitals – that they would not go to any clinic, even for dentistry or for serious medical emergencies.

The intersex rights movement began with the work of Tiger Devore. Devore was born with hypospadias, meaning that the opening of his urethra was not on the tip of his penis, but farther down the shaft. (Hypospadias is estimated to occur in 1 in 200–250 live male births[12] and is believed by some to be increasing in frequency due to environmental toxins including certain pharmaceuticals and pesticides.[13]) As was and still is "standard of care," a surgeon tried to create a new urinary opening at the tip of Devore's penis in infancy – not because there was any health problem with Devore's penis, but just so that Devore would eventually be able to urinate standing up like a "real" man.[14] Medical texts up until at least the 1990s insisted boys might grow up maladjusted and gay if they could not urinate in a steady stream while standing up.[15]

Hypospadias "correction" surgery is complicated because it requires building a urethral extension, and urethral tissue is very specialized and difficult to mimic. The surgery often leads to strictures, fistulas, and even necrosis (i.e., tissue death). A 2015 editorial in the *Journal of Urology* admitted: "For years we urologists routinely underestimated the problems our patients have experienced following hypospadias surgery."[16] The first penis "correcting" surgery failed in Tiger Devore's case. This led to another surgery, and another – ten by the time Devore was ten years old. By age fifty-five, he had had twenty-five genital surgeries to try to fix the problems set off by the first unsuccessful surgery. Medical texts call men like Devore "hypospadias cripples" because their penises are so damaged by repeated surgeries – surgeries done almost always because of utterly unproven psychological theories, not because they are medically necessary.

12 B. Troncoso and P.J. Lopez, "Hypospadias," in D. Wilcox, P. Godbole, and C. Cooper (eds.), *Pediatric Urology*, available at www.pediatricurologybook.com/hypospadias.html.

13 S. Botta, G.R. Cunha, and L.S. Baskin, "Do endocrine disruptors cause hypospadias?," *Translational Andrology and Urology*, 3:4 (2014), 330–339.

14 H. Devore, "Growing up in the surgical maelstrom," in A.D. Dreger (ed.), *Intersexuality in the Age of Ethics* (Frederick, MD: University Publishing Group, 1999), pp. 78–82.

15 A. Carmack, L. Notini, and B.D. Earp, "Should surgery for hypospadias be performed before an age of consent?," *Journal of Sex Research*, Epub ahead of print (October 19, 2015), 1–12.

16 D.A. Canning, "Can we correct hypospadias with a staged operation? If not, are we bold enough to report it?," *The Journal of Urology*, 194:2 (2015), 284–285.

Devore grew up to earn a PhD in clinical psychology and worked for a time at Johns Hopkins University with John Money. Not surprisingly for an out gay man in the 1970s–1980s in America, Devore came to develop a deep political consciousness about the clinical abuse of sexual minorities, and so he came to understand how intersex people, like gay people, were being harmed in the name of sexual norms. Because the literature on intersex made plain that the clinical goal of pediatric intersex treatment was not just stable gender identity but also the production of apparently heterosexual adults,[17] it was not hard for many gay intersex adults to see the political parallels and overlapping abuse. Devore began to appear as the first politically conscious out intersex person on national television programs, including on Oprah Winfrey's show in 1984. On these programs, he talked about his own experiences and the problems arising from the shame and secrecy that surrounds children born with uncommon forms of sex development.

Devore became known as someone to whom clinically traumatized intersex adults could be referred to for help. One of the people referred to him (in November 1992) was an intersex person then named Bonnie Sullivan, now named Bo Laurent. Sullivan/Laurent was born in 1956 with mixed sex anatomy, including ambiguous genitals (a phallic organ halfway between a penis and a clitoris), ovotestes (gonads that include ovarian and testicular tissue), and a vagina and uterus. First labeled a malformed boy (called Brian Sullivan), at eighteen months the child was surgically and socially reassigned to be a girl (and renamed Bonnie Sullivan). Surgeons removed all of her phallic tissue – every part of her penis/clitoris that they could excise – and later in her childhood they removed the testicular portions of her gonads to reduce a high risk of cancer and death. Never told by her doctors or parents what surgeries she had had, she grew up to be a lesbian who discovered she was missing a clitoris and missing sexual function.[18] Eventually she figured out what had happened and, with Devore's help, developed a political consciousness about it. As part of that, she eventually changed her name to Bo Laurent, with "Bo" being an androgynous form of "Bonnie," and

[17] See epilogue in A.D. Dreger, *Hermaphrodites and the Medical Invention of Sex* (Cambridge, MA: Harvard University Press, 1998), pp. 167–202.

[18] E. Weil, "What if it's (sort of) a boy and (sort of) a girl?," *New York Times* (September 24, 2006). Available at: www.nytimes.com/2006/09/24/magazine/24intersexkids.html. Accessed on: May 24, 2016.

"Laurent" being a tribute to Laurent Clerc, considered the father of Deaf political identity in America.

In response to Laurent's challenge of "give me something to live for or I'll commit suicide," Devore told me, "I taught her about the mistaken standard of care, the practices at Hopkins [and elsewhere], and the plight of so many Intersex people who had contacted me in the previous 10 years. I taught her about the value of mutual support groups, and suggested that she use her expertise" with computer technology to create a network of activist intersex people, "something that I wanted to do, and didn't know how to do."[19] In 1993, in response to an article by academic Anne Fausto-Sterling,[20] Sullivan/Laurent announced the founding of the Intersex Society of North America (ISNA).[21] For that activist work, she took the public activist name Cheryl Chase.

Devore contributed to ISNA and to the intersex rights movement the outline of what a reformed clinical system would look like: Intersex children would be given preliminary gender assignments as boys and girls (recognizing that all gender assignment is preliminary and does not require surgery); hormonal and surgical interventions would be limited to those that were needed to treat clear and present medical problems, with all elective interventions waiting until patients could consent for themselves; intersex children and adults (and their loved ones) would be provided professional, non-shaming psychosocial support and peer support.

Right around the time ISNA formed, when I was in my mid-twenties and working on a PhD in history and philosophy of science at Indiana University, I learned for the first time that humans can be born intersex. Knowing I was a feminist interested in gender and science, my dissertation director suggested I look at the history of hermaphroditism, particularly in medicine. In response to this suggestion, I consulted the nineteenth-century precursor to MedLine and, to my shock, found hundreds of journal articles documenting cases of children and adults with various sex anomalies. I decided to focus my dissertation research on the late nineteenth and early twentieth centuries in France and Britain, analyzing how doctors and surgeons made sense of about 300 cases that

[19] Personal email communication from Tiger Devore to Alice Dreger (May 14, 2016); used with permission.

[20] A. Fausto-Sterling, "The five sexes: Why male and female are not enough," *The Sciences*, 33:2 (1993), 20–25.

[21] See C. Chase, "Hermaphrodites with attitude."

came before them. Odd as this may sound, it did not really occur to me then to think about how there might be people all around me with the same conditions I was studying historically. At the time, I thought of intersex as a historical curiosity.

In 1995, I published my first work on the subject in the journal *Victorian Studies*.[22] This led Cheryl Chase/Bo Laurent to contact me via email to ask me what else I could tell her about the history of intersex, beyond what I had published.[23] (She explained to me why she had two names, one public and one private.) As she did with other academics who had wandered into the subject of intersex, she asked me to help change the current medical system, about which I knew nothing. I used my romantic partner's medical textbooks – he was in medical school at the time – as a starting point, and was stunned to find that "normal" modern-day care included hypospadias surgeries done to get boys to urinate standing up as an alleged means of preventing homosexuality and clitoral reduction surgeries done to prevent girls from growing up to be lesbians.

Through a series of conversations, Bo Laurent pulled me into the intersex rights movement. (I had been raised by activist anti-abortion parents, and so was familiar with the idea of helping with injustices. Unlike my conservative parents, however, my politics are progressive.) Laurent was soft-spoken and socially awkward, and yet somehow very good at getting people to help her. I think part of her success came from asking people point-blank for specific work. A task she gave me was helping intersex people (who counted as "lost to follow-up" in the medical literature) to record their histories in writing. That way their stories could be shared and we could help them learn to tell their stories coherently for the media.

I took to doing what I came to call "facilitated autobiographies." I would tape-record individuals telling me their stories, and then compose first-person autobiographical essays that the subjects could then correct as needed. Some people, like Tiger Devore, did not need this kind of help, but plenty did. These people's stories had strong themes of being actively lied to about their histories, of being estranged from their parents, of having had tissue removed that they wished had not been removed, of iatrogenic (medically caused) permanent tissue damage, and

22 A.D. Dreger, "Doubtful sex: The fate of the hermaphrodite in Victorian medicine," *Victorian Studies*, 38:3 (1995), 336–370.

23 See A.D. Dreger, "Cultural history and social activism."

of profound difficulty with sexual relationships. A few had changed gender labels – although none of those found the gender label revisions traumatic; what they found traumatic was the way they had been treated as children. When I collected some of these stories into the anthology *Intersex in the Age of Ethics* and I asked the authors to give me photos of themselves (in order to show a human face on intersex), it was impossible not to notice how many appeared with domesticated animals in their chosen photos.[24] This, I realized, was because so many had taken refuge in emotional relationships with pets following their trauma.

For years, everyone except my husband, Aron Sousa (the medical student I had been living with when I began my research into the history of hermaphroditism, now an internist), assumed that I was going to come out as a lesbian – that that was the only way to explain my passion about intersex rights. But if I had to guess at where my passion for the cause came from, I would say it arose primarily from a personality defect – being unable to give up when someone says I'm wrong about something when I am sure I am right. (This seems to be a common trait among whistleblowers.) I also felt a sibling-like devotion to Laurent; we had both been raised in sex-shaming Roman Catholic families with four children, and it was easy for me to imagine that I could have been her. I also appreciated her strong intellect and her strong interest in science. Given our mutual educational background in the sciences, she and I were both appalled to learn just how little science went into most medical care, and we pushed hard to try to get more science in intersex care, whether that be studies of prenatal androgens on ultimate gender identities or outcome studies on particular surgeries.

As Devore noted above, Laurent's professional background was in computers, and from the founding of ISNA forward, she used those skills to develop a powerful website (something fairly novel at the time) and to recruit allies. I was good at research, writing, editing, and speaking, so that was where I concentrated my efforts. Because I was not intersex and had a PhD, I could be and was used as an "expert" talking head in media accounts. Frustratingly, even though I was saying the same things as people like Devore and Laurent, reporters and producers generally counted Devore and Laurent only as intersex subjects, while I counted as a supposedly objective expert.

24 A.D. Dreger (ed.), *Intersex in the Age of Ethics* (Hagerstown, MD: University Publishing Group, 1999). Many of the essays in this volume were originally published in a special journal issue edited by A.D. Dreger, *Journal of Clinical Ethics*, 9:4 (1998), 337–420.

My position as a historian was useful to the intersex rights movement in a few other ways besides documenting individual histories and speaking to the media as an expert on the overarching history. For example, the work I had done showing what happened to intersex people before the era when surgeries became commonplace provided comfort to some living intersex people. One person told me that my work made her feel like she had discovered her ancestors, and many told me they cherished the photographs I reproduced of those "ancestors" from old medical journals. I was also able to help intersex people understand how their experiences were the result of cultural histories in which their doctors and parents had been like cogs in a cruel machine. This often helped them find some degree of peace or forgiveness.

Additionally, because I was a non-intersex historian, I was often seen as a "safe" person to invite to medical groups interested in hearing about intersex, and so it was often easier for me to get into the halls of medicine than intersex people who were feared as "angry former patients." As well (and perhaps more importantly), the history I documented showed that *the clinical approach needn't have been what it was*. This provided some measure of support for our collective activist belief that the health care system could be made different and better. What had come about in terms of clinical practice had not been inevitable. There was a time when intersex people decided for themselves, as adults, if they wanted surgeries; we could go "back" to this time.

I had graduated with my PhD in 1995 and then worked on converting my dissertation into a book published in 1998 by Harvard University Press.[25] In the book, I tackled the modern-day treatment system in a final chapter. That chapter was shared by the publisher in manuscript form with a pediatric endocrinologist at Harvard's medical school who participated in the system I was criticizing. He ruled that it was weakly supported in its claims that the current system was motivated by homophobia and sexism and that the current system was harmful. In response, I did more research prior to publication, strengthening the chapter considerably. When the revised chapter was presented to him, he said he could now see evidence for why the system had to change.

With the permission of Harvard University Press, I adapted the chapter about the current-day treatment system and intersex rights to become an article for the *Hastings Center Report*, then the leading North

[25] See A.D. Dreger, *Hermaphrodites and the Medical Invention of Sex*.

American bioethics journal. This article became the first published, sustained, academic ethical criticism of the modern-day treatment system.[26]

The Hastings Center publication excited those of us in the intersex rights movement, and I naively thought it would lead to some progress toward reform in the medical realm. It took me way too many years to learn that doctors generally don't really care what PhDs have to say in ethics journals unless the PhDs are supporting what the doctors already think. Laurent and I did try to get our criticisms published in medical journals, but we often ran up against rejection, except for short letters to editors. Meanwhile doctors writing in those journals repeatedly misrepresented ISNA's position, claiming we wanted to see children raised as hermaphrodites. They could not seem to understand that children could be raised as boys and girls without doing genital surgery, even though I had documented a long history of just that before the modern era of surgery. It was deeply frustrating to watch as misrepresentations of our positions came to be repeated, without physicians bothering to look up what we had written or to ask us what we really thought. It has taken me far too long to realize that the way to change doctors is not via academic or medical publications. Change is most effectively accomplished via the media – via perceived cultural pressure, especially the carrot of public heroism or the stick of public shaming – and via personal one-on-one relationships.

By the late 1990s, Laurent took ISNA firmly in the direction of advocacy rather than primarily peer support. I went along. In the early days, ISNA, like many other patient groups, had attempted to provide both active peer support and medical advocacy. This created tensions within ISNA, however, as has occurred in other medical reform groups I have observed. The problem is that the people who need peer support are usually still dealing with shame and are not out, whereas the people who do advocacy have moved on to pride and *are* out. Those members who are out grow impatient with those who are not.

ISNA was not by any means the only intersex policy-focused group, but while it existed (1993–2008), it tended to dominate the public scene. There was plenty of infighting in the intersex rights movement over the years, as within any activist movement.[27] As someone who was not

[26] A.D. Dreger, "'Ambiguous sex' – or ambivalent medicine? Ethical issues in the medical treatment of intersexuality," *Hastings Center Report*, 28:3 (1998), 24–35.

[27] K. Karkazis, *Fixing Sex: Intersex, Medical Authority, and Lived Experience* (Durham, NC: Duke University Press, 2008).

intersex, I took a risk choosing to ignore some intersex people who wanted to criticize and argue with me – I did not have an "identity card" to shield me, as identity cards often do in identity politics movements – but I generally took the risk of ignoring these critics because it seemed more important to spend my energy engaging with the people with power (i.e., physicians).

Around 1998, Laurent decided to formally incorporate ISNA as a federally recognized nonprofit organization. Because my husband and I had a stable address, ISNA was legally registered to our Michigan home, which remained the case for about seven years. I served as the first chair of the board. My husband Aron Sousa, Laurent's romantic partner Robin Mathias, and pediatric endocrinologist Bruce Wilson made up the rest of the board. While nonprofit incorporation made it easier to solicit donations and grants, we always operated on a shoestring budget. In addition to fundraising, I wrote mainstream and academic articles, worked openly and privately as an editor for writing that supported our work, and made a lot of media appearances. I also gave many talks in academic and nonacademic settings and provided a listening ear to clinicians who were themselves somewhat traumatized by what they had seen or participated in.

Unlike some of our fellow activists, Laurent and I made a point of learning all the medicine and science we needed to know in order to be taken seriously by physicians and other clinicians. It was common for hostile physicians to ask me to prove I knew one condition from another, knew the biochemistry, and so on. After I passed these tests, they would engage. (I was given bonus points for being married to a doctor, something I took to using as a joke at the start of my talks: "I'm not a doctor, but I sleep with one."[28])

Physicians would typically tell us that we were wrong to attribute the problematic treatment system to homophobia and sexism. Yet their own publications and presentations made it very plain that they were afraid intersex children would grow up gay. These same publications and presentations also made it very clear that these physicians believed fertility was a central goal for girls, and sexual performance was a central goal for boys. They told us that the "truth" was that parents could not bond with children with atypical sex and that children would be bullied into suicide. Physician-ally Bill Reiner and I tried repeatedly

[28] A.D. Dreger, "Sleeping with the enmity," *Atrium*, 3 (Fall 2006), 12.

to find any documented case of any intersex person in the modern era going "uncorrected" who had committed suicide. We could not find a single case. Notably, in over twenty years, not a single intersex person has come out publicly to say they are glad their parents elected early surgery for them. I think this shows how inadvertently shaming medical interventions cause more harm than good.

I frequently helped physician-allies write and rewrite papers, although I almost always kept my name off these papers. Today I feel ambivalent about the decision not to be listed as an author, not because I need the credit but because it now seems to me that this practice was to some degree dishonest. I suspect my participation in ghostwriting – where a person other than the author is given credit for the written work – suggests the extent to which I was thinking of much of this work as advocacy rather than scholarship. That so many physician-allies I wrote with never even suggested I get coauthorship also speaks to the degree to which they (and I) saw my work as "handmaiden" work.

My prominence in the intersex rights movement caused a few intersex activists to take potshots at me, claiming I was an interloper or a "wannabe," but most intersex people I worked with accepted me and indicated they appreciated the help. Laurent tended to try to focus ISNA on whatever her own current psychological and personal struggle required, but I tried to listen to other intersex people working with ISNA, to make sure the organization did not leave out the concerns of people who had had experiences different from Laurent's. So, for example, David Cameron Strachen, who has Klinefelter syndrome (XXY), helped me remember to talk not only about medically unnecessary surgeries but also about medically unnecessary hormone treatments that should wait until patients could be fully informed and decide for themselves.

Laurent and I both started to burn out after a few years, and we recognized that ISNA was sufficiently successful that it needed to professionalize – to have a real paid staff, a medical advisory board, and so on. We built up a war chest of cash and hired an academic sociologist to take over as executive director. The organization relocated from Laurent's home in northern California to Seattle (although legally it remained registered to my home address in Michigan). Unfortunately, that executive director left the organization after a few months. ISNA had thus suffered "founder's effect," the problem of when small organizations fail to survive when their founders depart. Laurent came back in and we refocused our efforts to try to stave off exhaustion. In 2004, I gave up my tenured professorship at Michigan State University, in part to focus on patient advocacy, although largely because I had grown exhausted trying to help run ISNA while simultaneously having a full-time job, a growing mainstream writing career, and raising a son born to me in 2000. (I was one of the few leaders in the intersex rights movement to have a child, something that again set me apart in that it really helped me to understand at a visceral level the intense and often self-defeating instinct to protect one's child at all cost.)

Around 2004, Laurent, our coworkers, and I obtained grant money for ISNA to produce two badly needed texts, namely, a set of reform clinical guidelines and a parents' handbook. I led these projects, and it was the hardest organizational work I have ever done. I used drafts produced years earlier by two social workers, Sallie Foley and Christine Feick, and substantially revised them with the help of more than fifty "consultants" from the three stakeholder groups: clinicians in the field, intersex adults, and parents of intersex children (young and grown). There were many differences of opinion I had to smooth out in many phone conversations, and a lot of time was spent educating people about why we were saying what we were saying. For the parents' handbook, intersex adults contributed essays on the theme of "what I wish my doctors had known," and parents contributed essays written as if they were letters to other parents. We included photos of intersex adults along with their childhood photos, so that people could see intersex children as grown up – to get doctors and parents to think about the long term.[29]

By this time (2004), the term "intersex" had become a heavily politicized term allied with the lesbian, gay, bisexual, and transgender (LGBT)[30] rights movement. This presented a problem. Pediatricians did not want to ascribe the heavily politicized (essentially queer) identity of "intersex" to babies, so among themselves they continued to use terms "true hermaphrodite" and "pseudo-hermaphrodite." But the "hermaphrodite" labels were understood by doctors to be stigmatizing enough that they often would still not tell parents these diagnoses, lest they upset them. Additionally, some adults who could logically be categorized as

[29] A.D. Dreger (ed.), *Clinical Guidelines for the Management of Disorders of Sex Development in Childhood* (Rohnert Park, CA: Intersex Society of North America, 2006). Available at: .dsdguidelines.org/node/11. Accessed on: May 25, 2016. A.D. Dreger (ed.), *Handbook for Parents: Consortium on the Management of Disorders of Sex Development* (Rohnert Park, CA: Intersex Society of North America, 2006). Available at: www.dsdguidelines.org/node/11. Accessed on: May 25, 2016.

[30] This is how the acronym appeared in circulation at the time. It has since changed to LGBTQ ("Q" standing for "questioning").

"intersex" from a biological standpoint rejected that term because they (reasonably) saw their conditions as medical conditions, not as political identities. Because the terminology remained contentious, physicians would often tell us that we were not even talking about the same conditions – supposedly they were talking about hypospadias or androgen insensitivity syndrome when we were talking about intersex. (Laurent and I believed that we were talking about the same things as the clinicians and that *they* were being evasive via semantic games.)

Laurent and I decided the two handbooks – the clinical guidelines and the parents' handbook I was editing under the grant support – would be a good place to launch a new terminology that might help physicians and parents see that intersex conditions could be understood as just that – conditions like other anatomical and medical conditions. We hoped such a change of vision would promote sorting out the medical from the psychosocial, with each treated appropriately. I penned an article for the *Journal of Pediatric Endocrinology and Metabolism*, coauthored by Laurent, Sousa, and allied physicians Philip Gruppuso and Joel Frader, laying out the reasons for a new nomenclature.[31] Fishing around for a new term, we suggested "disorders of sexual differentiation," which was already being used by some physicians. Ultimately, for the two handbooks we settled on "disorders of sex development" (DSD), which more clinicians preferred. As someone with a solid education in the politics of language who had spent years trying to demedicalize the treatment of intersex, this move felt jarring. Yet, like Laurent, I felt like we were stuck and could not move forward without pushing doctors to a new way of talking (and thus thinking), and it had to be a term that they would accept.

Around the same time, the major pediatric endocrinology societies held a "consensus workshop" in Chicago to respond to the intersex rights movement's ongoing demands for treatment reform. Laurent was invited, as was one other "intersex representative." Several allied physicians were also invited. I was not invited. I drafted "talking points" for those on our side of the treatment paradigm shift to help ensure a clear and consistent message as participants broke into smaller groups at the meeting to discuss topics that we expected would later be addressed in the larger consensus group – a kind of guerrilla marketing approach.

31 A.D. Dreger, C. Chase, A. Sousa, P.A. Gruppuso, and J. Frader, "Changing the nomenclature/taxonomy for intersex: A scientific and clinical rationale," *Journal of Pediatric Endocrinology & Metabolism*, 18:8 (2005), 729–733.

It worked; the "Chicago Consensus" included a statement that patients ending up gay or transgender should not be understood as treatment failures, that the overarching nomenclature should be changed from "hermaphroditism" to (persons with) "disorders of sex development," that patients and families should be told the truth about diagnoses and medical histories, that more outcomes data were badly needed, and that care should be done via teams including professional psychologists and social workers.[32]

The language of DSD caused controversy among activists who saw Laurent and I as "selling out" to a model of pathologizing intersex people, but we saw immediate positive side-effects in terms of clinicians now acknowledging that many different types of sex anomalies were all being (badly) "treated" because of cultural phobias around sex.[33] Additionally, many parents found it much easier to talk about "my child's DSD" than to talk about hermaphroditism or intersex – which meant they could talk about it.

Ultimately, however, the critics proved to be correct. The shift in terminology has not led to clinicians backing off on medically unnecessary surgeries. Indeed, constructing the conditions as "disorders" has fed some surgeons' attitude that "all disorders that can be corrected should be." I have lately joined those trying to shift the first "D" in DSD from "disorders" to "differences." For example, when I coedited a medical education guide for the Association of American Colleges on LGBT and DSD, at my recommendation we used "differences of sex development" as a way to shift away from the pathologizing terminology.[34] In this article I'm using "intersex" consciously since I am writing about the intersex rights movement, and I still often use that term, although in medical talks I tend to speak of "people with differences of sex development" – a "person-first" language that mimics the language of the disability rights movement.

[32] I.A. Hughes, C. Houk, S.F. Ahmed, P.A. Lee, and Lawson Wilkins Pediatric Endocrine Society/European Society for Paediatric Endocrinology Consensus Group, "Consensus statement on management of intersex disorders," *Journal of Pediatric Urology*, 2:3 (2006), 148–162.

[33] For some analyses of the politics of the terminology, see the following: E.K. Feder, "Normalizing medicine: Between 'intersexuals' and individuals with 'Disorders of Sex Development,'" *Health Care Analysis*, 17:2 (June 2009), 134–143. G. Davis, *Contesting Intersex: The Dubious Diagnosis* (New York, NY: NYU Press, 2015).

[34] A.D. Hollenbach, K.L. Eckstrand, and A. Dreger (eds.), *Implementing Curricular and Institutional Climate Changes to Improve Health Care for Individuals Who Are LGBT, Gender Nonconforming, or Born with DSD* (Washington, DC: Association of American Medical Colleges, 2014). Available at: http://members.aamc.org/eweb/upload/LGBTDSD%20Publication.pdf. Accessed on: May 25, 2016.

I reluctantly quit ISNA in 2005, just after the clinical guidelines and parents' handbook were published and made freely available online. Laurent and I had started knocking heads too frequently, disagreeing on management of projects, and I was tired of doing so much work and, frankly, getting so little credit. Once I quit ISNA, I became aware of how many other people had left ISNA without really wanting to, and how much work they had done that I had not been aware of. Many contacted me once I left ISNA and told me stories that suddenly sounded so similar to my own. I have since learned that movements like this retain people not simply based on merit or productivity, but rather on strength of personality. Most of us who felt we had to leave ISNA did not talk openly about this because we decided "the cause is the cause" – that it would be wrong to detract from the movement's goals by complaining about what happened within the organization. I have now come to realize, however, just how much work was done in and for ISNA, without my realizing it, by people like Tiger Devore, Sherri Groveman Morris, Emi Koyama, Kiira Triea, David Cameron Strachan, Thea Hillman, and others.

After quitting ISNA, I backed off intersex work for a while. But in 2009 I was called back in by old allies, including activist Janet Green, psychologist David Sandberg, and pediatric geneticist Eric Vilain, who by that point were running Accord Alliance – the successor organization to ISNA. They aimed to work from within medicine to change it. By then Laurent had closed ISNA and transferred its assets, including the clinical guidelines and the parents' handbook, to this new organization. The old friends contacting me wanted my help in trying to stop a specific problem: Pregnant women at risk of having a daughter with a particular inherited intersex condition were being given a development-altering steroid, dexamethasone, in an attempt to halt intersex development. The chief proponent of this off-label use of prenatal dexamethasone, pediatric endocrinologist Maria New, was telling prospective parents that it had been "found safe for mother and child" while at the same time obtaining National Institutes of Health research grants to use the same families retrospectively to see if it really was safe for mother and child. In the book *Galileo's Middle Finger*, I have told in detail the story of what happened – of how, with philosopher Ellen Feder, I organized scholars in bioethics and allied fields to try to mobilize government protections and investigations, and how and why we failed – so I will not tell that story again here.[35]

[35] A. Dreger, *Galileo's Middle Finger: Heretics, Activists, and One Scholar's Search for Justice* (New York: Penguin Press, 2015).

More recently I served as the consulting ethicist for Accord Alliance's National Institutes of Health–funded "Translational Research Network," a project aimed at supporting hospital-based pediatric teams seeking to move toward the reform model of care for DSD. The reform model is just what Devore had laid out, and ISNA had pursued, twenty years before: assigning a best-guess gender as boy or girl; holding off on elective procedures until the patient can consent for themself; providing rich psychosocial support, including peer support; treating medical dangers appropriately; and disclosing the full truth of diagnoses to parents and patients.

While many clinicians within the Translational Research Network have the right attitudes, practice has gone largely unchanged with regard to genital surgeries.[36] In 2010, after about a year of work, I resigned from the Network, along with Tiger Devore, who had also been an advisor. In our letters of resignation, we named the frustration of being twenty years in and still being told that parents have a right to order their children's genitals to be cut for social reasons.[37]

On most days, it is hard for me to face just how much I have failed to bring about real change in this scene. Some people have had more success; the lawyers who have come in to help, in particular Anne Tamar-Mattis, along with a new generation of intersex activists (and some of the old ones), have managed to elicit important statements from human rights groups about how what we are talking about when we talk about unconsented, medically unnecessary genital surgeries are human rights abuses. But so far those human rights statements have not changed practice.

Clinicians practicing the old way say that they and surgery-consenting parents are simply being realistic about our culture: You can't live well with a body that doesn't fit norms for male or female.[38] If this is true, it is

[36] S. Reardon, "The spectrum of sex development: Eric Vilain and the intersex controversy," *Nature*, 533:7602 (May 10, 2016), 160–163. Available at: www.nature.com/news/the-spectrum-of-sex-development-eric-vilain-and-the-intersex-controversy-1.19873. Accessed on: May 25, 2016.

[37] A. Dreger, "Rejecting the tranquilizing drug of gradualism in intersex care," Alice Domurat Dreger website (November 21, 2015). Available at: http://alicedreger.com/DSD_human_rights. Accessed on: May 25, 2016. A. Dreger, "Tiger Devore's statement," Alice Domurat Dreger website (November 21, 2015). Available at: http://alicedreger.com/Tiger_Devore. Accessed on: May 25, 2016.

[38] E.K. Feder, *Making Sense of Intersex: Changing Ethical Perspectives in Biomedicine* (Bloomington, IN: Indiana University Press, 2014).

going to take changing the culture to change medicine. That is what I work on mostly nowadays – mainstream writing aimed at getting people to know intersex exists, that it is just about natural developmental variations (some of which require medical attention for health reasons), that it does not have to be frightening, and that it should not exempt you from the legal, ethical, and moral protections other people enjoy automatically. As a historian, however, I know that before we see real change, the old guard has to die off. How very useless and impatient that makes me feel.

5

Working with Public Citizen

An Academic-NGO Collaboration

RUTH MACKLIN

If not for the committed advocacy work of Public Citizen's Health Research Group (hereafter "Public Citizen"), unethical aspects of research with human subjects may never come to light. I have had the privilege of working with this organization since 1999, in minor ways as well as in an ongoing activity more recently. Working with Public Citizen has been one way of fulfilling my initial purpose in entering the field of bioethics in its infancy in the early 1970s. That purpose was to engage in work that may make a difference in the world. Although I loved philosophy, the discipline in which I received my doctorate and taught for a few years, I always recalled the words of my thesis advisor, Isaac Levi: "Philosophy bakes no bread." I was interested in baking some bread, but not only in my own kitchen.

Still, in the early years my work in bioethics was exclusively academic. When I began teaching bioethics in 1971 along with several colleagues in the philosophy department at Case Western Reserve University, it was basically no different from academic work in philosophy: writing articles for publication, coediting the first anthology in the field,[1] and teaching students. After moving to the Hastings Center in the mid-1970s, I began to have a taste of what engagement in the world outside academia could be like. The Hastings Center was, and still is, a research institution unaffiliated with a university. It is not an advocacy organization like Public Citizen, but it is more than a mere think tank (Dan Callahan, one of the two founders and past president of the Center, shunned the term "think tank" as a descriptor). What I gained from my five years on the professional staff at the Hastings Center, among other benefits, was the ability to engage with scholars and professionals from different

[1] S. Gorovitz, A. Jameton, R. Macklin, J.M. O'Connor, E.V. Perrin, B. Page St. Clair and S. Sherwin (eds.), *Moral Problems in Medicine* (Englewood Cliffs, NJ: Prentice-Hall, 1976).

disciplines and experience talking about bioethical issues to nonacademic audiences. After joining the faculty of Albert Einstein College of Medicine (in New York) in 1980, I remained active in some of the work of the Hastings Center, focused largely on leading controversies of the day such as the removal of life supports from patients, the use of in vitro fertilization and other reproductive technologies, and the growing field of genetic screening and prenatal diagnosis. Yet while actively debating these and other contemporary bioethical challenges, the Hastings Center has not been an activist organization in the sense of advocating for a particular position or specific public policy.

I first learned about the work of Public Citizen in 1984. The occasion was my twenty-fifth class reunion at Cornell University. I was invited to serve on a panel with two others from the class of '59 who had what the program organizers called "unusual occupations." One was a journalist serving as the Washington bureau chief of a newspaper and the other was Sidney M. Wolfe, MD, the cofounder and long-time director of Public Citizen's Health Research Group.

Fifteen years after Sid Wolfe and I first met at the Cornell reunion, I was invited to serve as senior consultant to the National Bioethics Advisory Commission (NBAC), the presidential commission appointed by Bill Clinton. My work was devoted to organizing, overseeing, and contributing to a report the Commission had begun on international collaborative research, which was eventually published under the title *Ethical and Policy Issues in International Research: Clinical Trials in Developing Countries.*[2] Like many other US federal commissions and boards, NBAC operated under the Federal Advisory Committee Act,[3] which mandates that meetings be open to the public with time allotted for comments from members of the public. Commissions such as NBAC typically reserve time during their meetings for invited presentations by experts in the field. Soon after assuming my position as senior consultant to NBAC, I paid a visit to the office of Public Citizen to meet with Sid Wolfe and his then deputy director, Peter Lurie, accompanied by Alice Page, an associate working with me on the NBAC report. Before recounting the reason for our visit and proceeding

[2] National Bioethics Advisory Commission, *Ethical and Policy Issues in International Research: Clinical Trials in Developing Countries* (Bethesda, MD: National Bioethics Advisory Commission Publications, 2001).

[3] Federal Advisory Committee Act, Pub. L. 92–463, Sec. 1, Oct. 6, 1972, 86 Stat. 770.

to describe my more recent work with Public Citizen, I should explain more fully the organization and what it does.

Public Citizen's Health Research Group

The organization's website describes Public Citizen's work as follows: "Public Citizen's health and safety work protects consumers by advocating for safer, more effective drugs, medical devices and other products; equitable health care services and more physician accountability; safer cars and trucks; and improvements in worker safety."[4] The website describes specific activities related to drugs, devices, and supplements:

> When we feel that the information provided to doctors or patients is misleading, insufficient, or just plain wrong, we do not hesitate to express an opinion. Our book *Worst Pills, Best Pills*, our monthly print newsletter and our Web site[5] have been at the forefront of the provision of information to patients. We also use the law, petitions, and letters to monitor the FDA and coax the agency into action. Our actions have contributed to 23 drugs being pulled off the market.[6]

Public Citizen also seeks to ensure that research involving human beings is conducted in an ethical manner. An example of this is action the organization took in 2001 after it found out about a proposed study to investigate a drug to treat premature infants.[7] Premature infants that require ventilator support can suffer lung damage unless their lungs are protected by a surfactant (which is not internally produced until the baby is nearer to full term). Synthetic and animal-derived surfactants had been approved by drug regulatory agencies and were then in use throughout the world in hospitals with state-of-the art neonatal intensive care units. A US drug company, Discovery Labs, proposed a study of its experimental product, Surfaxin.[8] In addition to study sites in Europe,

4 "Health and safety," Public Citizen (2016). Available at: www.citizen.org/hrg. Accessed on: March 29, 2016.

5 "Worst pills, best pills," Public Citizen (2016). Available at: www.worstpills.org. Accessed on: March 29, 2016.

6 "Drugs, devices and supplements," Public Citizen (2016). Available at: www.citizen.org/Page.aspx?pid=1249. Accessed on: March 29, 2016.

7 P. Lurie and S.M. Wolfe, "The developing world as the 'answer' to the dreams of pharmaceutical companies: The Surfaxin story," in J.V. Lavery, C. Grady, E.R. Wahl, and E.J. Emanuel (eds.), *Ethical Issues in International Biomedical Research* (Oxford: Oxford University Press, 2007), pp. 159–170.

8 P. Lurie and S.M. Wolfe, "The developing world."

the company deliberately sought out locations in Latin America where premature newborns did not receive any surfactant (surfactants are expensive and therefore not available in some hospitals in resource-poor countries). In those countries, the company wanted to use a placebo control instead of, or in addition to, testing Surfaxin against a surfactant that was already approved by the Food and Drug Administration (FDA). A trial with a placebo control would have been considered unethical in the United States and Europe, where already approved surfactants were regularly used for infants with respiratory distress syndrome. Although the expectation was that Surfaxin, if proven safe and effective, would be marketed in the United States and Europe, the company seemed to have sought locations that did not have access to surfactants in order to justify a placebo-controlled trial.

At the same time that the placebo-controlled trial was to be carried out in hospitals in South America, Discovery Labs planned trials of their experimental product in Europe. The study design in the European countries compared Surfaxin with existing surfactants that had already been approved by drug regulatory agencies.[9] This is an example of the well-known double standard in biomedical research: one standard for the industrialized world and another – lower – standard for resource-poor countries.[10] Public Citizen obtained documents that were widely circulated at the FDA, which contained the revealing statement, "Conduct of a placebo-controlled surfactant trial for premature infants with RDS [respiratory distress syndrome] is considered unethical in the USA."[11] But the documents were ironically related to a scientific meeting at the FDA entitled "Use of Placebo Controls in Life Threatening Diseases: Is the Developing World the Answer?"[12] From the internal FDA documents obtained by Public Citizen, it appeared that the sponsor of Surfaxin was reluctant to design a trial that would compare its product with existing products for fear that the experimental product would not show at least similar efficacy.[13] Public Citizen's review of the literature revealed that no placebo-controlled trials of surfactants had been conducted since 1996,[14] but now

[9] P. Lurie and S.M. Wolfe, "The developing world."

[10] R. Macklin, *Double Standards in Medical Research in Developing Countries* (Cambridge: Cambridge University Press, 2004).

[11] P. Lurie and S.M. Wolfe, "The developing world."

[12] P. Lurie and S.M. Wolfe, "The developing world."

[13] P. Lurie and S.M. Wolfe, "The developing world."

[14] P. Lurie and S.M. Wolfe, "The developing world."

Discovery Labs was proposing one in locations where surfactants are not widely in use to treat newborns with respiratory distress syndrome.[15]

Taking its usual method of publicizing putatively unethical research and bringing it to the attention of relevant US governmental officials, Public Citizen wrote a letter to the then Secretary of the Department of Health and Human Services (HHS) describing the situation and asking that the Secretary take steps to ensure that the placebo-controlled trial would not be allowed to go forward.[16] The publicity produced by Public Citizen in this case had the desired result: The trials proposed for Bolivia and other Latin American countries were abandoned before they were begun.[17]

My Involvement with Public Citizen: The Beginning

The Surfaxin case was not the first time Public Citizen waged a campaign against placebo-controlled trials in developing countries. In a much more widely publicized episode, a group of US government–sponsored clinical trials that came to light in 1997 sparked a furious controversy that eventually led to the revision of major international[18] ethics guidance documents for research.[19]

The studies that Public Citizen openly criticized were mother-to-child HIV transmission studies carried out in Thailand, Côte d'Ivoire, and other countries in which some of the pregnant women were given a placebo even though a proven, effective treatment, azidothymidine (AZT), was available in industrialized countries.[20] The furor was initially prompted by an open letter addressed to US officials by Public Citizen,

[15] P. Lurie and S.M. Wolfe, "The developing world."

[16] P. Lurie and S.M. Wolfe, "The developing world."

[17] P. Lurie and S.M. Wolfe, "The developing world."

[18] P. Lurie and S.M. Wolfe, "Unethical trials of interventions to reduce perinatal transmission of the human immunodeficiency virus in developing countries," *New England Journal of Medicine*, 337:12 (1997), 853–856. H. Varmus and D. Satcher, "Ethical complexities of conducting research in developing countries," *New England Journal of Medicine*, 337:14 (1997): 1003–1005. Editorial, "The ethics industry," *The Lancet*, 350: 9082 (1997): 897. U. Schüklenk, "Unethical perinatal HIV transmission trials establish bad precedent," *Bioethics*, 12:4 (1998), 312–319. R. Macklin, "After Helsinki: Unresolved issues in international research," *Kennedy Institute of Ethics Journal*, 11:1 (2001), 17–36.

[19] R. Macklin, "After Helsinki."

[20] R. Macklin, "After Helsinki."

comparing the trials to the infamous Tuskegee experiments[21] and charging that the trials violated a clause in the Declaration of Helsinki,[22] the highly regarded international guidance for ethics in research. Public Citizen argued that an already proven treatment regimen, using AZT, could reduce the rate of mother-to-child transmission of the AIDS virus, so it was unethical to withhold that treatment from any women in the trial.[23] The international collaborative studies were carried out in countries that could not afford the expensive, high-dose AZT regimen routinely used in the United States and Europe. These clinical trials were testing a lower, shorter dose of AZT, which was much cheaper and therefore presumed to be affordable to the poorer countries that would make it available to pregnant women. The developing country studies also allowed AZT treatment much later in pregnancy, since women in those parts of the world do not routinely receive early prenatal care. And the drug was administered by mouth rather than through a vein, which was more in line with the medical facilities used in these countries. These departures from the proven treatment available in industrialized countries were intended to adapt the treatment to the medical realities in the developing countries where it could be introduced.

Once it had been shown to be effective in reducing the transmission of HIV from mother to child, the expensive AZT treatment became the standard therapy for HIV-positive pregnant women in the United States and other industrialized countries.[24] But it was argued that since women in the developing countries did not have access to any treatment whatsoever, they would not be made worse off by participating in the study than they would otherwise have been.[25] As a result of widespread publicity that followed the actions of Public Citizen, an extended public debate ensued. Some of the discussions appeared in the pages of leading academic and professional journals and involved the sponsors of the research – the National Institutes of Health (NIH), the Centers for Disease Control and Prevention (CDC), and the Joint United Nations Programme on HIV/AIDS (UNAIDS). Most notable were two dueling

21 CDC, "U.S. public health service syphilis study at Tuskegee," Centers for Disease Control and Prevention (December 14, 2015). Available at: www.cdc.gov/tuskegee/index.html. Accessed on: March 30, 2016.

22 WMA, *Declaration of Helsinki – Ethical Principles for Medical Research Involving Human Subjects*, World Medication Association (June 1964; amended October 2013). Available at: www.wma.net/en/30publications/10policies/b3/. Accessed on: March 30, 2016.

23 P. Lurie and S.M. Wolfe, "Unethical trials."

24 R. Macklin, "After Helsinki."

25 R. Macklin, "After Helsinki."

articles published two weeks apart in the *New England Journal of Medicine* (*NEJM*). The first was by Peter Lurie and Sid Wolfe criticizing the trials,[26] and the second was by Harold Varmus, then the director of the NIH, and David Satcher, then the director of the CDC[27] – the two main sponsors of the research. In their article, Lurie and Wolfe appropriately dismissed a claim used by defenders of the research that since the "standard of care" in the developing countries was "no treatment," participants in the placebo arm of the study would not be made worse off than if they had not been enrolled in the study. This would not be the last time Public Citizen became embroiled in a controversy in which their opponents invoked the notion of "standard of care" in defense of their position.

When NBAC was appointed by President Clinton, the controversy over the maternal-to-child HIV transmission research had not abated. The Commission decided that one of its reports should be devoted to international collaborative research.[28] Once that decision was made, the Commission's chair invited me to come on board as a senior consultant for the international project. That was when I paid a visit to the office of Public Citizen, and along with my colleague at NBAC, Alice Page, had a lively meeting with Sid Wolfe and Peter Lurie. They gave us each a copy of Public Citizen's book, *Worst Pills, Best Pills*,[29] which has sold more than two million copies. We inquired about their willingness to provide testimony at a meeting of NBAC, which they agreed to and subsequently did. I had found their arguments about the ethics of placebo controls persuasive, and the counterarguments unpersuasive. The bioethics community was divided on that issue, as it would later be on the more recent controversy in which I worked in a more sustained way with Public Citizen beginning in 2013.

The SUPPORT Controversy: Working with Public Citizen

Like many academics, I often learn about questionably ethical practices in medicine and research only when stories appear in the *New York*

[26] P. Lurie and S.M. Wolfe, "Unethical trials."
[27] H. Varmus and D. Satcher, "Ethical complexities."
[28] National Bioethics Advisory Commission, *Ethical and Policy Issues.*
[29] S.M. Wolfe, R-E. Hope, P.D. Stolley, and the Public Citizen Health Research Group, *Worst Pills, Best Pills II: The Older Adult's Guide to Avoiding Drug-Induced Death or Illness: 119 Pills You Should Not Use: 245 Safer Alternatives* (New York, NY: Pantheon Books, 1993).

Times or in an editorial in *NEJM* or the *Lancet*. This was the case for a controversy that erupted over the adequacy of informed consent documents in a study of treatment of severely premature newborns known as SUPPORT (the acronym stands for Surfactant, Positive Pressure, and Pulse Oximetry Randomized Trial). On April 10, 2013, an article in the *New York Times*[30] reported that a federal oversight agency had determined that consent forms in the SUPPORT study failed to disclose important information about risks to the extremely premature infants in the research. The article had been prompted by a publicly released letter sent on April 10, 2013, from Public Citizen to Kathleen Sebelius, then the Secretary of the HHS, criticizing the Office for Human Research Protections (OHRP) for not taking adequate actions in response to the agency's findings regarding the SUPPORT study.[31] On the same date, Public Citizen issued a press release describing the situation.[32]

I read the *New York Times* article with great interest, and a week later contributed a post to the Albert Einstein College of Medicine's blog, *The Doctor's Tablet*, endorsing the action by the federal agency, the OHRP.[33] Based on the brief *New York Times* article, my blog included the following points:

> Twenty three academic medical centers participated in the study of premature newborns, including prominent institutions like Stanford, Duke, and Yale. Why didn't any of the IRBs that reviewed the study notice that risks were not properly described in the consent documents? Is it because none of the reviewers on these committees read the consent forms carefully? Is it because IRBs tend to concentrate on simplifying the inordinately complex technical jargon in which forms are typically written?[34]

30 S. Tavernise, "Study of babies did not disclose risks, U.S. finds," *International New York Times* (April 10, 2013). Available at: www.nytimes.com/2013/04/11/health/parents-of-preemies-werent-told-of-risks-in-study.html. Accessed on: March 29, 2016.

31 M.A. Carome and S.M. Wolfe, "Letter to Secretary Sebelius – RE: The surfactant, positive pressure, and oxygenation randomized trial (SUPPORT)," Public Citizen (April 10, 2013). Available at: www.citizen.org/documents/2111.pdf. Accessed on: March 29, 2016.

32 "Press release – HHS-funded experiment exposed babies to risk of death and blindness without informing parents," Public Citizen (April 10, 2013). Available at: www.citizen.org/pressroom/pressroomredirect.cfm?ID=3859. Accessed on: March 29, 2016.

33 R. Macklin, "Informed consent in infant research: Ethical problems remain," *The Doctor's Tablet Blog* (April 18, 2013). Available at: http://blogs.einstein.yu.edu/informed-consent-in-research-ethical-problems-remain. Accessed on: March 29, 2016.

34 R. Macklin, "Informed consent in infant research."

And:

> One would have to read the entire [OHRP] 14-page letter (I did) to understand fully the charges, as well as the researchers' claim that infants in the study received the "standard of care" and that therefore, the study was "minimal risk." Nevertheless, that claim points to two controversial areas in the review of human subjects research. The first is uncertainty about how to determine what is the "standard of care" in medical practice. The second is whether a study should be considered "minimal risk" when the interventions carry significant risk but patients receiving similar interventions outside the study would undergo similar risks.[35]

These two controversial points were the basis for a furious, ongoing debate in medical and bioethics journals[36] over the adequacy of the consent forms in the SUPPORT study and, more broadly, how risks should be described in consent forms for research on the comparative effectiveness of treatments currently used in medical practice.

In retrospect, I believe that basing an ethical conclusion on a single newspaper article written by a journalist was a mistake. I knew nothing about the research in question other than what appeared in the article, and I had not seen the consent forms for the study that OHRP found lacking. Coincidentally, however, less than two weeks after my blog post appeared, Sid Wolfe contacted me asking if I would be willing to sign another letter about the SUPPORT study that Public Citizen was preparing to send to Secretary Sebelius. This second letter was drafted after Public Citizen had obtained more information and documents related to the study. The only other signatories on the letter I was invited to sign were Sidney M. Wolfe, MD, then still the director of Public Citizen's Health Research group, and Michael A. Carome, MD, then the deputy director (in the ensuing months Sid Wolfe stepped down and became senior advisor to Public Citizen, and Mike Carome became the new director). Sid believed it would strengthen Public Citizen's letter if a

[35] R. Macklin, "Informed consent in infant research."

[36] D. Magnus and A.L. Caplan, "Risk, consent, and SUPPORT," *New England Journal of Medicine*, 368:20 (2013), 1864–1865. J.M. Drazen, C.G. Solomon, and M.F. Greene, "Informed consent and SUPPORT," *New England Journal of Medicine*, 368:20 (2013), 1929–1931. B.S. Wilfond, D. Magnus, A.H. Antommaria, P. Appelbaum, J. Aschner, K.J. Barrington, et al., "The OHRP and SUPPORT," *New England Journal of Medicine*, 368:25 (2013), e36. R. Macklin, L. Shepherd, A. Dreger, A. Asch, F. Baylis, H. Brody, et al., "The OHRP and SUPPORT – Another view," *New England Journal of Medicine*, 369:2 (2013), e3. R. Macklin and L. Shepherd, "Informed consent and standard of care: What must be disclosed," *American Journal of Bioethics*, 13:12 (2013): 9–13.

bioethicist signed on, especially given growing interest in the controversy. In the intervening weeks since the *New York Times* article had appeared, *NEJM* published an editorial coauthored by the editor-in-chief[37] and an opinion piece by two leading bioethicists, David Magnus and Arthur Caplan.[38] Both articles criticized OHRP's action and defended the importance and ethical soundness of the SUPPORT study. Having already read the two *NEJM* articles on the opposing side of the ethical debate when I received Sid's letter on May 1, I was tentatively willing to sign Public Citizen's letter to the HHS Secretary. Sid sent me several background documents, some of which were obtained from government sources through the Freedom of Information Act. After reading through these documents, which included the original informed consent form approved by the Institutional Review Board (IRB) at the University of Alabama at Birmingham,[39] the lead institution in the multicenter SUPPORT study, I was eager to sign the letter to Secretary Sebelius. The letter I signed was dated May 8, 2013.[40]

What ensued thereafter was a full-blown controversy, with more articles appearing in journals and my continuing collaboration with Public Citizen's Health Research Group, along with other academics (Alice Dreger, Lois Shepherd, Leonard Glantz, Rebecca Dresser, Nancy M.P. King, and many others) who were critical of the informed consent documents in the SUPPORT study. Sid Wolfe's successor, Mike Carome, continued to provide important information about ongoing research similar to the SUPPORT study. Additional academics critical of the SUPPORT study kept up communications via email, which led to further published articles[41] that expanded the debate. Without the advocacy role of Public Citizen and its ability to provide critical updates, documents

[37] J.M. Drazen et al., "Informed consent and SUPPORT."

[38] D. Magnus and A.L. Caplan, "Risk, consent, and SUPPORT."

[39] This consent form is not a published document. It was obtained by Public Citizen through the Freedom of Information Act, 5 U.S.C. 552(a)(3), Section 10(b) of the Federal Advisory Committee Act (FACA), as amended, (Public Law 92–463, 5 U.S.C. App.).

[40] M.A. Carome, S.M. Wolfe and R. Macklin, "Letter to Secretary Sebelius – RE: The surfactant, positive pressure, and oxygenation randomized trial (SUPPORT) – Analysis of the complete protocol and complete consent form," Public Citizen (May 8, 2013). Available at: www.citizen.org/documents/2124.pdf. Accessed on: March 29, 2016.

[41] R. Macklin and L. Shepherd, "Informed consent and standard of care." W.D. Rich, K.J. Auten, M.G. Gantz, E.C. Hale, A.M. Hensman, N.S. Newman, et al. Antenatal consent in the SUPPORT trial: Challenges, costs, and representative enrollment," *Pediatrics*, 126:1 (2010), e215–e221; originally published online June 29, 2010; DOI: 10.1542/peds.2009-3353.

obtained through the Freedom of Information Act, and other resources, I might have languished in the ivory tower along with most of my fellow academics, lacking the timely information needed to contribute to emerging ethical controversies in medicine and health research.

SUPPORT: The Story Continues

The story of my involvement in the SUPPORT controversy might have ended with my agreement to sign the letter from Public Citizen to the Secretary of HHS, but when a letter to the editor appeared online on June 5, 2013, in the *NEJM*, signed by forty-six scholars in bioethics and pediatrics,[42] it reignited my interest. This letter again defended the informed consent documents in the SUPPORT study, acknowledging that the consent forms could have been improved, but went farther, claiming that the OHRP "overreached" in imposing a sanction (albeit a very mild one) on the lead institution conducting the study. The publication of yet another defense of the informed consent forms in the SUPPORT study deserved a response, if only another group of bioethicists could join forces.

An added benefit to having been invited to join Public Citizen's critique of the consent forms in the SUPPORT study was an introduction to two colleagues with whom I had not previously worked and who became allies in the cause. Public Citizen put me in touch with Alice Dreger, a historian and well-known contributor to ethics in action, who at the time was professor of clinical medical humanities and bioethics at the Feinberg School of Medicine at Northwestern University, and Lois Shepherd, professor of public health sciences and professor of law at the University of Virginia. Working together, the three of us began to round up bioethicists and other scholars willing to sign a letter to the *NEJM* editor countering the claims made by the forty-six bioethicists in the latest salvo. Public Citizen helpfully provided us with materials they had gathered through the Freedom of Information Act process, which included all of the consent forms used by the collaborating institutions in the multicenter SUPPORT study. In the end, our letter countering the one by the first group of bioethicists was published in the *NEJM* and had forty-five signatories.[43] Throughout the process of drafting and redrafting our article, we sought the advice of Mike Carome at Public

42 B.S. Wilfond et al., "The OHRP and SUPPORT."
43 R. Macklin et al., "The OHRP and SUPPORT – Another view."

Citizen, who was unfailingly helpful. He and Sid Wolfe graciously sent us a congratulatory email message when our letter appeared online on June 26, 2013.

The Struggle to Publish

Getting our one-page letter published in the *NEJM* was no easy task. The first hurdle was what to call the piece. It seemed appropriate to seek publication in the same category as the letter by Benjamin Wilfond, the lead author, and the forty-five others who criticized OHRP for "overreaching." That letter appeared in the journal under the heading "Correspondence." As we were still working on a draft of our article, we wondered whether to submit a commentary for the Perspectives section in which opinion pieces are published. That would enable us to write a longer piece than the length allowed for a letter. After a brief discussion, we decided that a Perspectives article would be more difficult to get accepted, so we opted for a letter similar to the one written by Wilfond et al.

On June 13, 2013, I wrote the following letter of inquiry to the editor of the *NEJM*:[44]

> Dear Dr. Drazen,
>
> We have drafted a letter to the editor in response to the Commentary by Benjamin S. Wilfond, which was signed by a group of bioethicists and pediatricians and published on line June 5, 2013. We note that instructions for letters to the editor say that "Letters in reference to a *Journal* article must not exceed 175 words (excluding references) . . . Letters not related to a *Journal* article must not exceed 400 words (excluding references)." Since our letter is related to a previously published letter to the editor, and not to a *Journal* article, we are uncertain whether our letter may exceed the 175-word limit. We note further that the Wilfond et al. letter contained 528 words. Therefore, we respectfully request that our letter, with a word count equivalent to that of the published letter, be considered for publication.
>
> The instructions also say that a letter can be signed by no more than three authors. In addition to Dr. Wilfond, who signed the published letter, 45 other names appear at the end of the letter. We have obtained an equivalent number of signatures from prominent bioethicists, physicians, law professors, and scholars in allied fields. Although the undersigned are the three authors of our letter, the additional signatures represent

[44] Correspondence from R. Macklin to J. Drazen (June 13, 2013). On file with the author.

> individuals who endorse the contents of the letter. They have all declared no conflict of interest.
>
> We would be grateful for a response to this query so that we may submit our letter within the three-week period for responses.
>
> Ruth Macklin
> Lois Shepherd
> Alice Dreger

Later the same day the editor, Dr. Drazen, replied:[45]

> Dear Professor Macklin,
>
> We consider the Wilfond letter to be a publication; our policy concerning a 175 word limit for letters applies.
>
> Best,
> jmd

After consulting with Alice and Lois, I wrote back thanking Dr. Drazen and said, "We will submit our article as a publication. Ruth Macklin."[46]

His reply said:[47]

> Dear Professor Macklin,
>
> Just to be sure I was clear. Your letter to the editor will be about the Wilfond letter and limited to 175 words.
>
> We look forward to your submission.
>
> Best,
> Jmd

This was not acceptable. Now what to do? We consulted again. Lois wrote: "Their letter starts 'To the editor' and is published under 'Correspondence.' So, it seems we don't care what they call it, we just want them to publish it."[48] Alice added: "I think Lois is right – tell him we'd like to know how theirs was categorized and we'll take that category."[49]

The next day Alice, Lois, and I signed and sent the following letter to Dr. Drazen:[50]

45 Correspondence from J. Drazen to R. Macklin (June 13, 2013). On file with the author.
46 Correspondence from R. Macklin to J. Drazen (June 13, 2013). On file with the author.
47 Correspondence from J. Drazen to R. Macklin (June 13, 2013). On file with the author.
48 Correspondence from L. Shepherd to R. Macklin (June 13, 2013). On file with the author.
49 Correspondence from A. Dreger to R. Macklin (June 13, 2013). On file with the author.
50 Correspondence from R. Macklin, L. Shepherd, and A. Dreger to J. Drazen (June 14, 2013). On file with the author.

Dear Dr. Drazen,

We have considered your reply to our query about publication. We are rather troubled by your response to our request to publish our view (and that of the many other bioethicists, physicians, and other scholars who have agreed to join us as signatories) with equal space given to the initial Wilfond letter. You mentioned that the Wilfond letter signed and published in the *NEJM* was not, in fact a letter. Because it was published under the heading "Correspondence," was addressed "To the Editor," and contains an equivalent wordcount and number of signatories, we do not see how it differs materially from our letter, other than that ours represents a differing opinion.

We would like to give you another chance to accept our letter. If we do not publish it with *NEJM*, we will certainly publish it elsewhere. Your denial of the request for equal space can only be seen as a suppression of viewpoint in light of the following: that your journal published the original SUPPORT study results, an editorial defending the informed consent documents, the Wilfond letter, and a commentary by NIH officials defending the study's consent process, all without significant publication of contrary viewpoints.

We hope you will reconsider your initial decision regarding publication of our letter.

Ruth Macklin
Lois Shepherd
Alice Dreger

Here is the editor's same-day reply:[51]

Dear Dr. Macklin,

We consider the Wilfond piece an article. Our policies our [*sic*] firm on this matter.

Best,
Jmd

After this response, we went back and forth trying to decide what the next step should be. We agreed that the best strategy would be to simply ask outright under what category the Wilfond article was published and request that our piece be published in the same category. There was a catch, however. As already noted, a letter to the editor must be within 175 words and may refer to articles published in *NEJM*. However, an article (or letter) in the 400-word category may not have references to articles previously published in the journal. This latter restriction meant that we could not refer to the Wilfond article – at least not directly – in

[51] Correspondence from J. Drazen to R. Macklin (June 14, 2013). On file with the author.

our piece. After considerable strategizing, we decided to let the matter rest over the weekend and on Monday morning, June 17, I sent the following email message to Dr. Drazen:[52]

> Dear Dr. Drazen,
>
> We've become aware there may be a misunderstanding in our communications. We propose to submit a general letter ("article") agreeing with the OHRP's determination that the SUPPORT consent documents were flawed. It will not reference a prior publication by the Journal. It will be under 400 words. The only remaining question about submission, then, would be whether the Journal will include the 40+ signatures of a number of prominent bioethicists, physicians, and law professors who agree with the contents of the article (letter). We assume that will not be a problem, since this was permitted for the Wilfond article. Please advise as to this last point, as we would like to move forward with submission soon.
>
> Thank you.
> Ruth Macklin
> Lois Shepherd
> Alice Dreger

His reply (the same day):[53]

> Dear Professor Macklin:
>
> Your proposal is acceptable.

After we received this reply, Lois reminded us that the reply did not constitute acceptance of our letter, but was simply an acknowledgment of our proposal. Throughout this process we consulted Mike Carome at Public Citizen, who was always helpful with questions like this: "Have the 'negotiations' involved discussions of whether your letter would be published online, like all of the other *NEJM* commentaries/letters/editorials on this issue so far?"[54] I had to acknowledge that we had not thought about that.

On June 20 we received an email accepting our piece, worded as follows: "I am pleased to inform you that your Letter to the Editor entitled, 'The OHRP and SUPPORT: Another view,' has been accepted for online-only publication in edited form at NEJM.org."[55] Thinking this

[52] Correspondence from R. Macklin to J. Drazen (June 17, 2013). On file with the author.
[53] Correspondence from J. Drazen to R. Macklin (June 17, 2013). On file with the author.
[54] Correspondence from M. Carome to R Macklin (June 18, 2013). On file with the author.
[55] Correspondence from Edward W. Campion, M.D. Senior Deputy Editor and Online Editor to R. Macklin (June 20, 2013). On file with the author.

was the end of the process, we made sure that everyone who signed the letter submitted conflict of interest forms and consent to publish. I received a PDF file of the article in the form it was to be published. To my surprise, in addition to the references we provided, none of which mentioned the articles on SUPPORT previously published in *NEJM*, there were now references to the editorial by Drazen and colleagues, and the other articles by Magnus and Caplan, Wilfond et al., and one by the director of the NIH and two senior NIH colleagues[56] – all of which had previously been published in the journal. This clearly violated the journal's own policy.

At this late stage I was dealing with the copy editor by telephone, cleaning up some details. When I pointed out the citations that were not in the manuscript we submitted and that we were precluded from including such references according to *NEJM* policy,[57] the copy editor replied that it was one of the top editors who insisted on putting in those citations.[58] I said we wanted to ensure that the editors know that we did not make those references, and she said that "it came from the very top."[59] I replied to the copy editor by email as follows: "My co-authors and I are quite disappointed about the addition of the four citations and the associated sentence. I know you said the decision was made by the top editors, but my co-authors believe it changes the content of what we wrote originally. We played by the rules, but it appears that the rules have changed."[60] The copy editor's reply: "I forwarded your message to the editor who had requested the addition. We have now deleted 'recently discussed in the *Journal*' and have deleted references 1 to 4 (leaving, now, only the former reference 5, which is now reference 1)."[61] As it turned out, the editors still could not resist referring in some manner to the previously published articles in *NEJM*. Using a journal editor's prerogative they added an "Editor's note" referencing all of the articles.

[56] K.L. Hudson, A.E. Guttmacher, and F.S. Collins, "In support of SUPPORT. A view from the NIH," *New England Journal of Medicine*, 368:25 (2013), 2249–2251.

[57] "Article types," *New England Journal of Medicine* (2016). Available at: www.nejm.org/page/author-center/article-types. Accessed on: October 17, 2016.

[58] Phone conversation between R. Macklin and *New England Journal of Medicine* copy editor (June 24, 2013).

[59] Phone conversation between R. Macklin and *New England Journal of Medicine* copy editor (June 24, 2013).

[60] Correspondence from R. Macklin to *New England Journal of Medicine* copy editor (June 24, 2013). On file with the author.

[61] Correspondence from *New England Journal of Medicine* copy editor to R. Macklin (June 24, 2013). On file with the author.

The Saga Continues

Our *NEJM* article was not to be the last word in the controversy. David Magnus, the editor of the *American Journal of Bioethics* (*AJOB*), who also coauthored with Arthur Caplan the early defense of SUPPORT in *NEJM*, invited a number of bioethicists to contribute to an issue of *AJOB* with further reflections on the controversy.[62] I was one of those invited to contribute, and with Lois Shepherd as coauthor, I wrote an article recounting the story so far and defending the action by OHRP.[63] We shared this work in progress with Mike Carome, who made helpful suggestions for improvement.

Following this flurry of activity, a brief hiatus ensued, only to be broken by announcement by the HHS at the end of June that a public meeting, prompted by the SUPPORT study controversy, had been scheduled for August 2013. Mike Carome informed us about this meeting and circulated the announcement entitled, "Notice of a Department of Health and Human Services Public Meeting and Request for Comments on Matters Related to the Protection of Human Subjects and Research Studying Standard of Care Interventions."[64] Individuals interested in attending and speaking at that meeting had to register with OHRP and indicate their wish to speak. I was disappointed that I was unable to participate, as I had made a previous commitment many months earlier to speak at a conference in Reykjavik, Iceland. However, Alice Dreger and Susan Reverby, one of the other signers of our *NEJM* letter, collaborated on a powerful statement in their public comment at the meeting, and Lois spoke persuasively, as well.

A Back Story

In its determination that the University of Alabama at Birmingham had failed to comply with the US federal regulations regarding what must appear in informed consent documents, OHRP imposed a very mild

[62] Special issue on the SUPPORT controversy. *American Journal of Bioethics*, 13:12 (2013).

[63] R. Macklin and L. Shepherd, "Informed consent and standard of care."

[64] Office of the Secretary, Department of Health and Human Services, "Notice of a Department of Health and Human Services public meeting and request for comments on matters related to the protection of human subjects and research studying standard of care interventions," Department of Health and Human Services (June 26, 2013). Available at: www.gpo.gov/fdsys/pkg/FR-2013-06-26/html/2013-15160.htm. Accessed on: September 28, 2016.

sanction. On page 11 of its thirteen-page letter to the University of Alabama at Birmingham on March 7, 2013, OHRP wrote:

> Accordingly, we determine that the informed consent document for this trial failed to adequately inform parents of the reasonably foreseeable risks and discomforts of research participation.
>
> UAB [University of Alabama at Birmingham] Required Actions: Please provide a plan that the IRB will use to ensure that approved informed consent documents include and adequately address the basic elements of consent as required by HHS regulations at 45 CFR 46.116(a) . . . Please provide responses to the above determinations by March 22, 2013, including a corrective action plan to address the determination.[65]

On June 4, 2013, OHRP wrote a second letter to University of Alabama at Birmingham withdrawing the sanction it had imposed in its earlier letter.[66] However, OHRP had not changed its mind about the failure of the consent documents to comply with the federal regulations. If anything, the second letter from OHRP strengthened its earlier criticisms. So, what had transpired in the period between OHRP's two letters to lead the agency to withdraw its sanction? One thing was evident: All the publicity generated by the articles in the *NEJM*, as well as the actions by Public Citizen, attracted the attention of journalists and bloggers. An article that appeared in *The Scientist* on June 10, 2013, included the following information: "In a June 4 letter to administrators at the University of Alabama, Birmingham, which led the so-called SUPPORT trial, the US Office for Human Research Protections (OHRP) stated that it would suspend its previous decision to discipline researchers involved in the study for failing to fully inform parents of the potential risks involved."[67]

What was not evident at the time was interference by top-ranking NIH officials in the retraction of sanctions by OHRP, which also involved other high officials in the HHS. Additional efforts by Public Citizen to

65 Office for Human Research Protections, "Letter to the University of Alabama at Birmingham - RE: Human research protections under federal wide assurance (FWA) 5960," Department of Health and Human Services (March 7, 2013). Available at: www.hhs.gov/ohrp/detrm_letrs/YR13/mar13a.pdf. Accessed on: March 29, 2016.

66 Office for Human Research Protections, "Letter to the University of Alabama at Birmingham - RE: Human research protections under federal wide assurance (FWA) 5960," Department of Health and Human Services (June 4, 2013). Available at: www.hhs.gov/ohrp/detrm_letrs/YR13/jun13a.pdf. Accessed on: March 29, 2016.

67 B. Grant, "Leaders of infant trial will not yet face sanctions," *The Scientist* (June 10, 2013). Available at: www.thescientist.com/?articles.view/articleNo/35918/title/Leaders-of-Infant-Trial-Will-Not-Yet-Face-Sanctions. Accessed on: March 29, 2016.

acquire internal documents from the NIH revealed deep collaboration between NIH officials, the HHS Office of the Secretary, and the agency that is supposed to oversee OHRP and compliance with federal regulations governing research. These communications and events are described below.

On May 20, 2014, slightly more than a year after the story of OHRP's actions on the SUPPORT study broke, Public Citizen wrote a letter to the HHS inspector general with the heading "Senior HHS Officials Allow NIH to Interfere with OHRP's Compliance Oversight Investigation of the Surfactant Positive Airway Pressure and Pulse Oximetry Trial in Extremely Low Birth Weight Infants: The SUPPORT Study."[68] The letter was posted on Public Citizen's website and requested that the inspector general:

> immediately open a formal investigation to scrutinize the dangerous and unacceptable conduct of senior leaders in the Department of Health and Human Services' (HHS's) immediate Office of the Secretary (OS), the Office of the Assistant Secretary for Health (OASH), and other OS components who knowingly allowed the director of the National Institutes of Health (NIH) and other senior NIH officials to interfere with the independence of the Office for Human Research Protections' (OHRP's) ongoing compliance oversight investigation of the unethical SUPPORT study.[69]

Public Citizen's letter contained details of correspondence between high-ranking NIH and HHS officials and Jerry Menikoff, the director of OHRP, obtained through the Freedom of Information Act. Although the numerous email exchanges were heavily redacted, there was sufficient information to demonstrate the efforts of NIH to pressure Menikoff to withdraw sanctions against the SUPPORT investigators. Public Citizen's letter alleged that the pressure exerted by NIH over an independent federal oversight agency amounted to a serious conflict of interest.

Among these email messages was one dated May 1, 2013, from Francis Collins, the NIH director, to Bill Corr, deputy secretary of HHS: "I have been closely tracking efforts by my staff (Kathy Hudson, Alan

[68] "Letter to Inspector General Levinson – RE: Senior HHS officials allow NIH to interfere with OHRP's compliance oversight investigation of the surfactant positive airway pressure and pulse oximetry trial in extremely low birth weight infants: The SUPPORT study," Public Citizen (May 20, 2014). Available at: www.citizen.org/documents/2201.pdf. Accessed on: March 29, 2016.

[69] "Letter to Inspector General Levinson," Public Citizen.

Guttmacher, and others) who have been working productively with Howard Koh and others in [OASH] and OHRP, to develop a consensus set of statements that OHRP could put forward to clarify the situation with the SUPPORT study."[70] Menikoff (the OHRP director) then sent a new letter he had drafted to several NIH and HHS officials,[71] and received a reply from Kathy Hudson, the NIH deputy director, with the subject line "Suggested correction to OHRP-UAB draft letter."[72] The back-and-forth among Menikoff, Hudson, Collins, and other HHS officials continued through June 4, 2013, with Collins again writing on June 2: "Thank you for the opportunity to weigh in on OHRP's letter to UAB [University of Alabama at Birmingham] and the Federal Register Notice related to SUPPORT. I have pasted NIH's comments on each of those documents below."[73] From Collins's email message dated June 4, it is evident that he had been in touch with the editor of *NEJM* about this matter, as he wrote: "*NEJM* has confirmed that they will post the NIH essay at 5 PM on Wednesday."[74] The next day, a Perspectives article was published in *NEJM* with the title "In support of SUPPORT – A view from the NIH."[75] The authors were Kathy Hudson, Francis Collins, and Alan Guttmacher, the head of the NIH institute that funded the SUPPORT study.

Mike Carome invited several of us who had collaborated on the letter published the previous year in *NEJM* to sign Public Citizen's letter to the HHS inspector general. Those of us who agreed added our signatures to those of Mike and Sid Wolfe: Larry Churchill, Alice Dreger, Rebecca Dresser, Carl Elliot, Nancy King, Susan Reverby, Lois Shepherd, Roy Spece, and I.[76] In addition, Congresswoman Rosa DeLauro wrote a letter

70 Available at Public Citizen's website: www.citizen.org/documents/support-study-HHS-internal-emails.pdf. Accessed on: October 17, 2016.

71 Available at Public Citizen's website: www.citizen.org/pressroom/pressroomredirect.cfm?ID=4184. Accessed on: October 17, 2016.

72 Available at Public Citizen's website: www.citizen.org/pressroom/pressroomredirect.cfm?ID=4184. Accessed on: October 17, 2016.

73 Available at Public Citizen's website: www.citizen.org/pressroom/pressroomredirect.cfm?ID=4184. Accessed on: October 17, 2016.

74 Available at Public Citizen's website: www.citizen.org/pressroom/pressroomredirect.cfm?ID=4184. Accessed on: October 17, 2016.

75 K.L. Hudson, A.E. Guttmacher and F.S. Collins, "In support of SUPPORT."

76 The letter is available on Public Citizen's website: www.citizen.org/documents/2201.pdf. Accessed on: October 17, 2016.

to the HHS inspector general on May 20 endorsing Public Citizen's call for an immediate formal investigation.[77]

Despite Public Citizen's allegation – and our agreement – that this matter constituted "meddling of NIH in OHRP's compliance oversight investigation,"[78] the HHS inspector general declined to take meaningful action. The HHS Office of the Inspector General issued a six-page report in September 2014.[79] The report included the following statements:

> Our review found no indication that NIH or other Department of HHS officials sought to intervene in OHRP's initial determination letter regarding SUPPORT. Our review did confirm that after the initial letter was issued, NIH encouraged OHRP to reverse its determination and that NIH and others in HHS provided input on OHRP's later clarification letter . . . Our research disclosed no law, regulation, or written policy that prohibits or restricts the kind of consultation that occurred here or would make such consultations improper . . . No law mandates OHRP's operational independence. Accordingly, we found no basis for action against the individuals involved.[80]

This legalistic response to the charge that NIH improperly meddled in the work of OHRP, the agency responsible for oversight of all HHS-funded research, blatantly ignores the ethics of the situation. It needs no reminder that not everything that is legally permissible is ethically acceptable. Although the published letter to the HHS Office of the Inspector General had attracted some attention in the media,[81] it failed to elicit a critical response from the only agency that might have acknowledged the wrongful interference by NIH and high officials in HHS.

The Aftermath

On October 24, 2014, OHRP posted on its website a document entitled "Draft Guidance on Disclosing Reasonably Foreseeable Risks in Research

[77] R.L. DeLauro, "Letter to Inspector General Levinson," United States House of Representatives (May 20, 2014). Available at: www.citizen.org/documents/2201b.pdf. Accessed on: March 29, 2016.

[78] Stated in the "Letter to Inspector General Levinson," Public Citizen.

[79] Office of Inspector General, *OIG Review of Allegations of Interference in OHRP Compliance Determination for SUPPORT*, Report OIG-12-14-04, U.S. Department of Health and Human Services (September 15, 2014). Available at: http://oig.hhs.gov/reports-and-publications/portfolio/oig-12-14-04.asp. -Accessed on: March 29, 2016.

[80] Office of Inspector General, *OIG Review of Allegations*.

[81] Media reports are on Public Citizen's website, available at: www.citizen.org/documents/support-study-HHS-internal-emails.pdf. Accessed on: October 17, 2016.

Evaluating Standards of Care," dated October 20.[82] Mike Carome circulated the draft guidance document as it appeared in the *Federal Register* notice the day before the post appeared on the OHRP website. As Mike noted in an email message to Alice, Lois, and me – the three of us who had worked together on the letter in *NEJM* and continued to follow the later developments – "OHRP overall appears to have come down on our side."[83] As is the case for draft guidance and proposed rule-making, the notice published in the *Federal Register* provided a sixty-day period for comments on OHRP's guidance. In a related development, the Institute of Medicine (now renamed the "National Academy of Medicine") announced "a one-and-a-half day workshop on December 2–3, 2014 to explore questions related to ethics and review of human subject protections in research trials studying standard of care interventions."[84] I was invited to serve on a panel at the workshop along with others on both sides of the controversy surrounding the SUPPORT study. There I met Mike Carome in person for the first time.

This workshop began with a video of Jerry Menikoff explaining the OHRP draft guidance and making frequent references to the SUPPORT study (in spite of an admonition to speakers not to refer specifically to that study). The positions taken by panelists were predictable. Panelists were instructed not to address the SUPPORT study specifically but to direct their remarks to the general issues raised by so-called standard of care research. Yet some individuals were unable to avoid reference to SUPPORT, referring to it as "the study that shall not be named." Some who had been among the most vocal defenders of SUPPORT intensified their defense, arguing – implausibly – that if consent forms had to be written in the way we critics insisted, "this type of research would grind to a halt."

Mike Carome had not been invited to be one of the panelists, but he spoke in a session devoted to comments from the public. The thrust of

[82] Office of Human Research Protections, *Draft Guidance on Disclosing Reasonably Foreseeable Risks in Research Evaluating Standards of Care*, U.S. Department of Health and Human Services (October 20, 2014). Available at: www.hhs.gov/ohrp/newsroom/rfc/comstdofcare.html. Accessed on: March 29, 2016.

[83] Correspondence from M. Carome to R. Macklin, L. Shepherd, and A. Dreger (October 23, 2014). On file with the author.

[84] *Ethical Review and Oversight Issues in Research Involving Standard of Care Interventions: A Workshop*, the National Academies of Sciences, Engineering, and Medicine (December 2–3, 2014). Available at: www.nationalacademies.org/hmd/Activities/Research/StandardofCare.aspx. Accessed on: October 17, 2016.

his presentation was to demonstrate that the SUPPORT study was decidedly not an example of "standard of care" research. Mike argued that research ethics committees (IRBs in the United States) need to carefully assess whether research that purports to be "usual care" does, in fact, meet that criterion. Saying that the draft OHRP guidance could be somewhat clearer, he basically endorsed the guidance as requiring disclosure to human research subjects of the type of risks that had been excluded from the SUPPORT study's consent form, a position Public Citizen and those of us who joined its efforts had strongly advocated since the SUPPORT study controversy erupted. The entire proceedings of the workshop, including Menikoff's video presentation on the OHRP draft guidance, are preserved on video.[85]

Conclusions

I found working with Public Citizen to be a most worthwhile effort and would be pleased to engage in future collaborative efforts. Among other lessons learned, it serves as a reminder of how difficult it is to challenge high-ranking officials, be they heads of governmental organizations or editors of prestigious medical journals. The successful effort of the NIH to get OHRP to rescind its disciplinary action against the University of Alabama at Birmingham had a special irony for me. That is because I served on a governmental review panel convened in 1999 to examine the relationship between the NIH and OHRP's predecessor, the Office of Protection from Research Risks.[86]

At the time, the Office of Protection from Research Risks was located *within* the NIH. The organization charged with oversight of federally funded research was therefore subordinate to one of the most important institutions whose research it was charged with overseeing for compliance with federal regulations. This arrangement was correctly perceived to be a conflict of interest, and the review panel on which I served

[85] Institute of Medicine, *Ethical Review and Oversight Issues in Research Involving Standard of Care Interventions: A Workshop* (video), the National Academies of Sciences, Engineering, and Medicine (December 2, 2014). Available at: www.nationalacademies.org/hmd/Activities/Research/StandardofCare/2014-DEC-02/Day%201/Opening%20and%20Session%20I/9-Public-Comment-Video.aspx. Accessed on: March 29, 2016.

[86] Report to the Advisory Committee to the Director, NIH from the Office for Protection from Research Risks Review Panel, National Institutes of Health (June 3, 1999). Available at: http://acd.od.nih.gov/reports/060399b_OPRR_Review_Panel.htm. Accessed on: October 17, 2016.

recommended removing the Office of Protection from Research Risks from the NIH and placing it in a location within HHS that could enable it to be independent of the NIH director and deputy directors. Harold Varmus, who was then the NIH director, and Donna Shalala, who was then Secretary of HHS, concurred with the recommendation and in 2000 the Office of Protection from Research Risks was removed from the NIH and replaced with the new organization, OHRP, placed under supervision of the assistant secretary for health in HHS. But as the correspondence obtained by Public Citizen clearly revealed, both Howard Koh, the assistant secretary for health, and Bill Corr, deputy secretary of HHS, were copied on various email exchanges between the NIH officials and Menikoff.[87] Formal independence of OHRP from NIH has apparently not had the intended effect of removing the oversight body from the powerful influence of the NIH.

[87] Available at Public Citizen's website: Available at: www.citizen.org/pressroom/pressroomredirect.cfm?ID=4184. Accessed on: October 17, 2016.

6

Reproductive Technology's Legacy of Omission

MIRIAM ZOLL

It was 2 a.m. on a cold winter night in 2006. I remember that distinctly. I had left the warm comfort of my bed and tiptoed downstairs, as I often did, to whimper like an injured animal into a pillow on the couch at the other end of the house.

This had become a pattern over the last number of months, this inability to sleep and an irrepressible need to cry. My husband and I were deep inside a three-year assisted reproductive technology (ART) odyssey for what the doctors told us was my "age-related infertility." Decades earlier, it would have been called the "onset of menopause." Our fourth in vitro fertilization (IVF) cycle had recently failed and our doctor was now recommending egg donation.

"We're just taking one cell from the egg donor but your blood and your amino acids will course through the baby's body, will help it to grow," the infertility clinic's social worker told us with a warm smile. "You won't be the genetic mother, but you will be the biological mother." As though that explanation made the idea of buying another woman's DNA appear as commonplace as a blood transfusion, she escorted us to the door where she advised us not to talk about our situation to people outside the clinic setting.

"*We* understand what you are going through," she said, patting my shoulder. "But other people might be judgmental, especially about donor eggs. Just be careful who you share that information with."

Like well-trained puppies, we did as she suggested and kept our ordeal and its excruciating decision-making processes concealed from public view. Sleepless nights ensued. Were we really going to try to make a baby by merging my beloved husband's sperm with some stranger's eggs that we bought online? I was a feminist, a journalist, and long-time health and rights advocate. I was the founding coproducer of the Ms. Foundation for Women's original *Take Our Daughters to Work Day*, and had worked for the United Nations and Planned Parenthood. It made no sense to me politically or philosophically, and yet I was compelled to try, in part because the medical professionals we trusted encouraged us to do so, but also because we were isolated and ill informed.

But the decision was also motivated by the intense remorse I felt for having waited so long to attempt childbearing. In the end, it was the guilt that propelled me to sign up for two so-called donor egg cycles with two different women. After detecting health problems in each of the donors, the doctors decided to terminate both cycles. The first woman had a genetic disease she did not know she had until we paid for tests that flagged it. The second woman produced a dozen eggs after potent hormone stimulation, but none of them fertilized with my husband's sperm.

"There is something obviously wrong with this second egg donor," the doctor told us over the phone the day before I was scheduled for an embryo transfer. "I would never recommend using her again."

I was speechless. It had never occurred to me that the donor's eggs might be compromised. We had not known to inquire about this possibility because we assumed the "egg broker" would have gone to great lengths to ensure that the women advertised on their website (for a hefty fee) were healthy and that their eggs – the "product" we were buying – were in working order.

To add insult to injury, while delivering this sixth batch of crushing news in thirty-six months, our doctor actually had the audacity to make one final sales pitch. As I recall, it went something like this: "I sure hope this negative outcome doesn't discourage you from trying another donor cycle. You are a perfect candidate."

That was the moment that substantiated my former suspicions that my husband and I had been snared in a trap – a profit-driven medical marketing trap, a cultural trap, a trap of misinformation; a trap that intentionally preyed upon the naiveté and vulnerability of anxious couples. It was also the moment I intuitively sensed that we were not the only ones investing enormous amounts of emotional energy and money in overrated medical services that frequently failed. The fact that our doctors had not sufficiently informed us about the limitations of reproductive technologies made me wonder just how transparent information sharing was between the infertility industry, the media, and general public.

During those dismal and depressing final months of treatment I had frantically searched for information online that might help me understand the trauma, confusion, and ethical dilemmas linked to the truly strange and often bizarre choices and circumstances we had been confronted with. In those days, before I learned what information to look for and where, all I was able to access were clinic websites featuring photos of smiling, predominantly white couples holding gurgling babies, and news headlines announcing one scientific "breakthrough" and "miracle birth" after another.[1] The industry rhetoric and cultural sentiment blanketing IVF was celebratory and hopeful. Yet I knew from my own experience that this superficial impression was not always the norm. Still, I could not quite identify what felt so out of alignment to me about the public face of repro tech.

I remember reading a pamphlet in the clinic waiting room when we first started treatments and I still deeply trusted the doctors and clinic staff. I was *thrilled* to see the words: "Sixty percent of patients become parents" printed in bold letters and I immediately decided that we would be one of the lucky ones. Just like that. Magic. I was too naïve to know that the statistics were skewed,[2] that I was now living inside the echo chamber of the repro tech industry's powerful, persuasive, and well-funded marketing narrative,[3] one that had driven public perception and policy discussion through a lens of "success" for decades.[4]

A case in point was our very first consultation with our reproductive endocrinologist. To his credit, he told us that given my age – I was forty-two – our chances with IVF were very slim. But, at the same time, during our fifteen-minute meeting with him, he did not present us with a folder of information documenting stubbornly high failure rates across *all* ages

[1] J. Hawkins, "U.S. Fertility care: Ignoring minorities?," *Huffington Post* (December 1, 2012). Available at: https://www.huffingtonpost.com/jim-hawkins/fertility-care-ignoring-m_b_2225069.html. Accessed on: February 20, 2018.

[2] M. Zoll, "The IVF data wars," *Biopolitical Times*, Center for Genetics and Society (November 15, 2013). Available at: www.biopoliticaltimes.org/article.php?id=7310. Accessed on: February 20, 2018.

[3] A. Preisler, "Assisted reproductive technology: The dangers of an unregulated market and the need for reform," *DePaul Journal of Health Care Law*, 5:2 (Fall 2013), 212–236. "Top 7 ways to market your fertility practice in 2014," Kaufer DMC. Available at: https://kauferdmc.com/blog/top-7-ways-to-market-your-fertility-practice-in-2014. Accessed on: February 20, 2018. I. Levitan, "The 2nd IVF worldwide live congress: A marketing wrap-up," *IVF Worldwide Blog*. Available at: https://ivf-worldwide.com/ivf-worldwide-blog/the-2nd-ivf-worldwide-live-congress-a-marketing-wrap-up.html. Accessed on: February 20, 2018.

[4] S. Dingle, "IVF doctors misleading women about success rates, industry experts say," *ABC* (May 29, 2016). Available at: http://www.abc.net.au/news/2016-05-30/ivf-doctors-misleading-patients-about-success-rates-experts-say/7457750. Accessed on: February 20, 2018. D. Rochman, "5 million babies born through IVF in past 35 years, researchers say," *NBC News* (October 14, 2013). Available at: www.nbcnews.com/health/5-million-babies-born-through-ivf-past-35-years-researchers-8C11390532. Accessed on: February 20, 2018.

that has plagued IVF attempts since 1978,[5] when the first "test-tube baby" was born in Britain.[6] He did not direct us to the Centers for Disease Control and Prevention (CDC) website[7] where we could review national data about ART or related health risks for women or potential offspring. We did not even know such a database existed. Nor did he share with us that in the best case scenarios IVF *might* be able to bump a couple's chances of conceiving up to the same rate they might conceive naturally. According to the American Society for Reproductive Medicine (ASRM), one of the leading trade and lobby groups for the US infertility industry, a healthy, fertile thirty-year-old woman has a 20 percent chance of conceiving during any given month, while a forty-year-old woman has less than a 5 percent chance. Among conception-challenged couples, IVF can sometimes match these rates.[8]

What our doctor did tell us was that no one could predict the outcome of an IVF cycle. While simultaneously wearing his medical hat and his business hat he said: "I've seen women your age and older birth children with their own eggs. You just never know."

And so there it was, hanging like a ripe mango ready to pluck from the tree, the infertility industry's seductive challenge: How hungry were we – and the millions of other hopeful couples around the world that walk into repro tech clinics every year – for a genetically related child? Did we have the collective tenacity to gamble (with the help of science), or would we retreat and choose instead adoption, or foster care, or even a life with no children? What were we made of? Little did I know that I was stepping into a surreal and life-altering experience that would later position me as an outspoken critic of the ART industry.

Shortly after our sixth ART attempt failed, I was serendipitously invited to join the board of Our Bodies Ourselves. Founded in the United States in the early 1970s, Our Bodies Ourselves was one of the world's first

[5] R. Edwards, "The bumpy road to human in vitro fertilization," *Nature Medicine*, 7:10 (October 2001), 1091–1094.

[6] "World's first test-tube baby born," *This Day in History* (1978). Available at: www.history.com/this-day-in-history/worlds-first-test-tube-baby-born. Accessed on: February 20, 2018.

[7] "Assisted reproductive technology (ART)," Centers for Disease Control and Prevention (2016). Available at: www.cdc.gov/art/index.html. Accessed on February 20, 2018.

[8] "Age and fertility: A guide for patients," American Society for Reproductive Medicine, Patient Information Series (2012). Available at: http://www.care4ba.com/uploads/Age_and_Fertility_ASRM.pdf. Accessed on: February 20, 2018.

women's health organizations.[9] Today it continues its tradition of providing objective health and human rights information to millions of women and girls in dozens of countries. It is one of the only women's health organizations in the United States that has no fiscal ties to the medical or pharmaceutical industries, relying instead on support from foundations and individuals.[10]

Much to my surprise, I learned that, like me, Our Bodies Ourselves was concerned about the global repro tech industry. Over the last several years it has raised important questions about health risks to women and infants due to multiple embryo transfers, and human rights concerns regarding cross-border commercial gestational surrogacy and egg selling.[11] Due to the absence of adequate medical evidence and long-term studies, Our Bodies Ourselves regards the infertility industry with caution and many of its services as experimental.[12]

My affiliation with them immediately connected me to a global grassroots repro tech watchdog network I had not even known existed. The network includes hundreds of organizations (including women's health groups, human rights organizations, government agencies, university think tanks, groups tracking genetic engineering, and religious organizations), as well as thousands of individuals (including physicians, pediatricians, bioethicists, academics, surrogates, egg donors, donor-conceived children, and former ART consumers and patients).

After my husband and I walked away from ART I was treated for post-traumatic stress disorder (PTSD). Interacting with and sharing information with members of this international network was vital to my recovery process and my personal growth, particularly when information they provided me with confirmed my own assessments and analysis of the industry. Network members became indispensable allies in my efforts, through journalism and public speaking engagements, to insert a much-needed consumer perspective into the public and media discourse about repro tech. The more

9 "History," Our Bodies Ourselves (2016). Available at: www.ourbodiesourselves.org/history/. Accessed on: http://www.care4ba.com/uploads/Age_and_Fertility_ASRM.pdf.

10 "About us," Our Bodies Ourselves (2016). Available at: www.ourbodiesourselves.org/about/. Accessed on: February 20, 2018.

11 "The ethics of assisted reproductive technologies," Our Bodies Ourselves (2016). Available at: www.ourbodiesourselves.org/health-info/the-ethics-of-art/. Accessed on: February 20, 2018.

12 J. Norsigian and M. Goodwin, "Popular fertility treatments still a vast experiment," *Women's e-news* (August 31, 2013). Available at: http://womensenews.org/2013/08/popular-fertility-treatments-still-vast-experiment/. Accessed on: February 20, 2018.

data I accumulated, the greater was my desire to share it with the public, and especially with reproductive age women and men. I began taking copious notes, reading numerous studies, and slowly, over time, I connected the dots – not just about my own personal experience, but about many of the complex and troubling long-term health and ethical implications of ART on individuals, families, and society.[13]

I had spent many months in isolation trying to make sense of ART and coping with its all-encompassing intrusion into my body, sexuality, and marriage. I was an astute and curious detective at heart, but my attempts to research ART online had rendered disappointing results. This is why I was so relieved when network members began flooding my inbox with background information on a range of issues I knew very little about.

This included human rights abuses of poor, barely literate women in countries like India, Mexico, and Thailand that signed up to be gestational surrogates because they desperately needed the cash,[14] and the increasing incidence of "extra" made-to-order IVF babies that were being trafficked on the black market.[15] Former so-called egg donors told me how their egg brokers profited from their reproductive tissue by selling their oocytes to six different couples instead of two, and about their hospitalizations and health challenges after hormonal hyperstimulation.[16]

Network members directed me to dozens of short-term investigations about the harm certain drugs and IVF techniques posed to both women.

[13] J. Johnston and M. Zoll, "Is freezing your eggs dangerous? A primer," *The New Republic* (November 1, 2014). Available at: www.newrepublic.com/article/120077/dangers-and-realities-egg-freezing. Accessed on: February 20, 2018. M. Zoll, "Higher risks for artificially conceived infants," *Center for Genetics and Society* (March 12, 2014). Available at: www.geneticsandsociety.org/article.php?id=7588. Accessed on: February 20, 2018. M. Zoll, "Hospital delivery charges significantly higher when babies conceived through assisted reproductive technologies," *Center for Health Journalism* (March 4, 2014). Available at: www.reportingonhealth.org/2014/03/03/hospital-delivery-charges-significantly-higher-babies-conceived-through-assisted. Accessed on: February 20, 2018.

[14] C. Vincent and A.D. Aftandilian, "Liberation or exploitation: Commercial surrogacy and the Indian surrogate," *Suffolk Transnational Law Review*, 36:3 (September 2013), 671–685. Available at: http://bit.ly/2epuGrI. Accessed on: February 20, 2018.

[15] S. Mohapatra, "Stateless babies and adoption scams: A bioethical analysis of international commercial surrogacy," *Berkeley J. Int'l L.*, 30:2 (2012), 412–450. Available at: http://bit.ly/2dTXGbV. Accessed on: February 20, 2018. K.D. Krawiec, "Altruism and intermediation in the market for babies," *Washington and Lee Law Review*, 66 (2009), 203–257. Available at: http://bit.ly/2eDOebb. Accessed on: February 20, 2018.

[16] "M" (pseudonym), "What I wish I knew before I donated my eggs," *We Are Egg Donors* (November 13, 2015). Available at: http://bit.ly/2f0J7Gl. Accessed on: February 20, 2018.

I learned about the potentially dangerous off-label use of drugs like Lupron[17] and possible connections between hormone stimulation and increased rates of ovarian, endometrial, or breast cancer.[18] I discovered that while my clinic had a patient and consumer cut-off age of fifty, many other clinics refused to treat women over the age of forty-two because the failure rates and emotional toll were so high.

I was grateful that so many trusted voices in women's health and human rights were validating my misgivings about this supposedly self-regulated[19] branch of medicine, and that many of my questions about both physical and emotional harms were being answered. At the same time, exposure to this new and very valuable information caused me to wonder why I hadn't known all this *before* I started infertility treatments. Why hadn't my doctors and their clinic staff shared these kinds of materials with my husband and me? And why hadn't I been able to locate more health research and academic studies on my own online?

One very important reason is that I did not know to look for them, but also because so many scholarly papers are locked behind pay walls. Often I could not access relevant data unless I paid hundreds of dollars in fees and subscriptions, and I opted not to do that. This decision undoubtedly inhibited my efforts during treatments to acquire information that would have been valuable to my husband and me as we waded through unprecedented and complex choices in a vacuum.

But I later discovered that another reason why information on the Internet is so often weighted toward IVF success stories and the hope of science has to do with the infertility industry itself. Unfortunately, it has

[17] J. Mroz, "High doses of hormones faulted in fertility care," *New York Times* (July 16, 2012). Available at: http://www.nytimes.com/2012/07/17/health/research/high-doses-of-hormones-add-to-ivf-complications.html. Accessed on: February 20, 2018. "Lupron: What does it do to women's health?," *National Women's Health Network Newsletter* (September/October 2008). Available at: www.nwhn.org/lupron-what-does-it-do-to-womens-health/. Accessed on: February 20, 2018.

[18] M.M. Reigstad, I.K. Larsen, T.Å. Myklebust, T.E. Robsahm, N.B. Oldereid, A.K. Omland, et al., "Risk of breast cancer following fertility treatment: A registry based cohort study of parous women in Norway," *International Journal of Cancer*, 136:5 (2015), 1140–1148. P. Russell, "IVF tied to higher ovarian cancer risk?," *WebMD Health News* (October 20, 2015). Available at: www.webmd.com/infertility-and-reproduction/news/20151020/ivf-ovarian-cancer-risk. Accessed on: February 20, 2018.

[19] M. Ollove, "States not eager to regulate fertility industry," *Stateline*, The PEW Charitable Trusts (March 18, 2015). Available at: http://bit.ly/1LxxWOQ. Accessed on: February 20, 2018.

evolved into a mostly profit-driven medical business[20] with an interest in guarding trade secrets and promoting a successful brand,[21] even if that means omitting details I believe patients and consumers have a right to know.

Indeed, a historical legacy of omission existed within this fledgling branch of medicine even before Louise Brown, the world's first IVF baby, was born in Britain in 1978.[22] A case in point is the early IVF experiments orchestrated at Oldham District and General Hospital by the medical team responsible for her birth – gynecologist Patrick Steptoe, physiologist Robert Edwards, and lab technician Jean Purdy.[23]

After Brown's much celebrated and sensationalized arrival on July 25, 1978,[24] the trio waited two years before publishing what was recently revealed to be only partial details of the procedures that led to her birth, and of almost a decade's worth of experiments performed on close to

[20] J. Wardell and J. Range, "Asia's low fertility trap opens opportunities in IVF Market," *Reuters* (June 13, 2103). Available at: https://www.reuters.com/article/us-asia-ivf/asias-low-fertility-trap-opens-opportunities-in-ivf-market-idUSBRE95C03K20130613. Accessed on: February 20, 2018. A. Picchi, "Why lenders see profits in fertilization treatments," *CBS Money Watch* (May 19, 2015). Available at: www.cbsnews.com/news/why-funding-fertility-is-a-growing-business/. Accessed on: February 20, 2018. K. Aubusson, "Baby-booming: The business of IVF," *Australian Doctor* (January 21, 2015). Available at: www.australiandoctor.com.au/news/news-review/baby-booming-the-business-of-ivf. Accessed on: February 20, 2018.

[21] M. Morris, "IVF industry criticized for 'misleading claims, aggressive marketing,'" *ABC News* (February 9, 2016). Available at: www.abc.net.au/news/2016-02-09/ivf-industry-criticised-over-'misleading-claims'/7152508. Accessed on: February 20, 2018. M. Zoll and P. Tsigdinos, "Selling the fantasy of fertility," *New York Times* (September 11, 2013). Available at: http://www.nytimes.com/2013/09/12/opinion/selling-the-fantasy-of-fertility.html. Accessed on: February 20, 2018. "Baby miracle begins at sandwich factory," news.com.au (November 29, 2013). Available at: http://www.news.com.au/lifestyle/parenting/baby-miracle-begins-at-sandwich-factory/news-story/c4b499b6a138caf95b67fe421f47baf7. Accessed on: February 20, 2018.

[22] M. Zoll, "Questioning the cult of repro tech," *The Dark Mountain Project* (October 21, 2015). Available at: http://dark-mountain.net/blog/questioning-the-cult-of-repro-tech/. Accessed on: February 20, 2018.

[23] K. Elder and M. Johnson, "*The Oldham Notebooks*: An analysis of the development of IVF 1969–1978. II. The treatment cycles and their outcomes," *Reproductive Biomedicine & Society Online*, 1:1 (June 2015), 9–18.

[24] "The IVF global experience," Australian Research Council Discovery Project, University of Wollongong. Available at: www.uowblogs.com/ivfglobalexperience/. Accessed on: February 20, 2018. "Test tube babies," *PBS International* (undated). Available at: http://pbsinternational.org/programs/test-tube-babies/. Accessed on: February 20, 2018.

three hundred or more working-class women[25] at Oldham Hospital, outside of Manchester, England.

The medical team's long-awaited IVF summary article, "Clinical aspects of pregnancies established with cleaving embryos grown *in vitro*," was published in the *British Journal of Obstetrics and Gynecology*[26] in 1980, the same year Steptoe and Edwards opened Bourn Hall, their for-profit clinic.[27] The article described 79 women who underwent 45 natural IVF cycles from November 9, 1977, to August 1, 1978.

But the recent discovery and 2015 publication of archival records, *The Oldham Notebooks*,[28] shows that at least an additional 203 research subjects submitted to the rigors of stimulated cycles and 389 laparoscopic egg retrieval surgeries from 1969 to 1978.[29] The data also reveal that 106 of these women submitted to surgeries (or multiple surgeries) and hormone regimens at a time when any resulting embryos were only being transferred into rabbit oviducts,[30] a revelation that raises serious questions about informed consent practices at Oldham Hospital at that time.

Additional information from *The Oldham Notebooks* indicates that the team covertly performed many of their experiments and procedures[31] and required participants to keep the experiments a secret from family and friends.[32] Noni Fallows, a nurse who worked at Oldham, says in an interview[33] that it was between 1970 and 1971 that Steptoe and Edwards

[25] K. Elder and M.H. Johnson, "*The Oldham Notebooks* II. The treatment cycles and their outcomes."

[26] P.C. Steptoe, R.G. Edwards and J. Purdy, "Clinical aspects of pregnancies established with cleaving embryos grown *in vitro*," *British Journal of Obstetrics and Gynaecology*, 87:9 (1980), 757–768.

[27] "Our family," Bourn Hall Fertility Clinic (2018). Available at: https://www.bournhall.co.uk/our-family/. Accessed on: February 20, 2018.

[28] K. Elder and M.H. Johnson, "*The Oldham Notebooks* II. The treatment cycles and their outcomes."

[29] K. Elder and M.H. Johnson, "*The Oldham Notebooks* II. The treatment cycles and their outcomes."

[30] K. Elder and M.H. Johnson, "*The Oldham Notebooks* II. The treatment cycles and their outcomes."

[31] K. Elder and M.H. Johnson, "*The Oldham Notebooks* II. The treatment cycles and their outcomes."

[32] M.H. Johnson, G. MacDonald, and K. Elder, "Supplementary material: Edited transcript of interview with Grace MacDonald conducted at Bourn Hall on 25 June 2013," Supplemental Materials attached to article by K. Elder and M.H. Johnson, "*The Oldham Notebooks* II. The treatment cycles and their outcomes." Available at: http://www.rbmsociety.com/article/S2405-6618(15)00003-9/addons. Accessed on: February 20, 2018.

[33] K. Elder, N. Fallows, M.H. Johnson, and J. Webster, "Supplementary Material 1: Edited transcript of interview with Noni Fallows and John Webster conducted at Bourn Hall on

moved their egg retrieval operations away from the main hospital campus to Kershaw's Cottage, about two miles away. She recalls Steptoe saying, "We're going to have a little theater set up away from here, Noni, and away from the hospital, somewhere secret, and we'll need good nurses in the theaters."[34]

It may well be that the 1980 ribbon cutting ceremony at Bourn Hall birthed the modern era of the *ART and IVF medical entrepreneur*,[35] the health professional that must be well versed in medicine, marketing, and business in order to competitively generate sales and attract patients.[36]

Based on my research and personal experience in the United States, where there is virtually no external oversight, patients are often referred to as out-of-pocket paying "customers" or "clients,"[37] and doctors and clinics are viewed as being more concerned with profits than the delivery of safe and appropriate health services.[38] As a result, a growing tension is mounting[39] between doctors that treat patients suffering from specific

19 May 2014," Supplemental Materials attached to M.H. Johnson and K. Elder, "*The Oldham Notebooks*: An analysis of the development of IVF 1969–1978. IV. Ethical aspects," *Reproductive Biomedicine & Society*, 1:1 (June 2015), 34–45. Available at: http://www.rbmsociety.com/article/S2405-6618(15)00003-9/addons. Accessed on: February 20, 2018.

34 K. Elder et al., "Supplementary Material 1: Edited transcript of interview with Noni Fallows and John Webster."

35 L. Skountridaki, "The internationalisation of healthcare and business aspirations of medical professionals," *Sociology*, 49:3 (June 2015), 471–487. Available at: http://journals.sagepub.com/doi/abs/10.1177/0038038514535863. Accessed on: February 20, 2018. C. Shalev and G. Werner-Felmayer, "Patterns of globalized reproduction: Egg cells regulation in Israel and Austria," *Israel Journal of Health Policy Research*, 1:15 (April 2012), 1–12.

36 G. Jones, "6 ways savvy fertility clinics are preparing to take IVF cycles from you in the next recession," *Fertility Bridge* (February 11, 2016). Available at: http://www.fertilitybridge.com/blog/6wayssavvyfertilityclinicswilltakeivfcyclesfromyouinthenextrecession. Accessed on: February 20, 2018.

37 "'My dear American clients': 8 tips on how to avoid turning off potential English-speaking clients," Global IVF: The #1 Free Resource Guide for Fertility Travel (2015). Available at: http://globalivf.com/2012/09/03/my-dear-american-clients/. Accessed on: February 20, 2018. G. Jones, "6 Ways Savvy Fertility Clinics."

38 M. Zoll, "Pregnant at 60: Where there's a uterus, there's a way?," *The Atlantic* (May 22, 2013). Available at: www.theatlantic.com/health/archive/2013/05/pregnant-at-60/276106/#comments. Accessed on: February 20, 2018. P. Ubel, "Are *in vitro* fertilization (IVF) clinics scamming customers?," *The Health Care Blog* (December 12, 2012). Available at: http://thehealthcareblog.com/blog/2012/12/16/are-in-vitro-fertilization-ivf-clinics-scamming-customers/. Accessed on: February 20, 2018.

39 L.J. Martin, "Anticipating infertility: Egg freezing, genetic preservation, and risk," *Gender & Society*, 24:4 (2010), 526–545. Available at: http://gas.sagepub.com/content/24/4/526. Accessed on: February 20, 2018.

reproductive maladies versus those providing expensive elective services to healthy women hoping to sell or freeze their eggs, or conquer the early phases of menopause through ART interventions.[40] As the respected medical ethicist George Annas has suggested, the behaviors of many for-profit clinics and affiliated businesses are, unfortunately, eroding the reputation of what was once a more well-regarded and trusted branch of medicine but is now often referred to as the "Wild West."[41]

In 2016, the global ART industry generated revenue of close to US$10 billion. It is anticipated that this will double to approximately US$20 billion by 2023. One of the driving forces of this market growth is increased use by older women.[42]

***

Buoyed and inspired by my role on the board of Our Bodies Ourselves and new affiliations with the global ART watchdog network, I decided to adapt three years' worth of diary entries into a book manuscript. I hoped that by telling my own story of coming of age with the concurrent promises of the women's movement and reproductive technologies that I could help ignite a long overdue discussion among patients/consumers and the general public about motherhood, career, and the role that science and technology plays in our reproductive lives. I felt very strongly that people needed to be warned about the high IVF failure rates[43] and potential health risks[44] that are so often not discussed in clinic settings or

[40] E.I. Kamphuis, S. Bhattacharya, F. van der Veen, B.W. Mol, and A.A. Templeton, "Are we overusing IVF?," *BMJ*, 348:252 (January 28, 2014), 1–5. Available at: www.bmj.com/content/348/bmj.g252. Accessed on: February 20, 2018.

[41] "Interview: George Annas," *Frontline*, PBS (undated). Available at: https://www.pbs.org/wgbh/pages/frontline/shows/fertility/interviews/annas.html. Accessed on: February 20, 2018.

[42] "In Vitro Fertilization (IVF) services market by cycle type (fresh, thawed IVF and donor egg IVF cycle) and end user (fertility clinics, hospitals, surgical centres and clinical research institutes) – Global opportunity analysis and Industry forecast, 2017–2023," PR Newswire and Allied Market Research (December 2017). Available at: https://www.alliedmarketresearch.com/IVF-in-vitro-fertilization-services-market. Accessed on: February 19, 2018.

[43] J. Brody, "The misleading promise of I.V.F. for women over 40," *New York Times* (October 17, 2016). Available at: http://nyti.ms/2eE7EgJ. Accessed on: February 20, 2018.

[44] J. Brody, "Some I.V.F. experts discourage multiple births," *New York Times* (October 10, 2016). Available at: http://nyti.ms/2dGOZCg. Accessed on: February 20, 2018.

in the media, and I wanted to candidly share insights about some of the very challenging emotional and ethical concerns that might arise for couples during infertility treatments.

I titled my memoir *Cracked Open: Liberty, Fertility and the Pursuit of High-Tech Babies.*[45] The book blends my experience as a repro tech consumer with my professional background as a journalist and health and rights advocate. It was not easy to weave these three aspects of my identity together, especially since I was still recuperating from the physical and emotional aftershocks of treatments. Sometimes it took longer than I expected to process and digest findings from studies that, for example, highlighted possible connections between IVF cycles, hormone exposure, and cancer.[46] There I was, wearing my reporter and advocacy hat, writing a book that I hoped would alert the public to the fact that very real health risks might exist for women, while at the same time, I was a former ART consumer who had directly exposed myself to the risks being discussed.

Many times I became so overwhelmed by new information that I resolved to stop writing. In those moments, I would place the manuscript on the shelf in the closet and close the door. But usually within a week or two another sensationalized headline with extremely misleading information would catch my eye, or someone would email me a study linking IVF to childhood leukemia,[47] for example, and I would feel morally obliged to keep writing. Words, after all, can be weapons of change.

This became apparent during the final editing stages of *Cracked Open* when two very important shifts occurred in my thinking and my sense of self. The first shift was linked to my personal evolution as a public speaker on this particular issue, where I was openly playing the role of David critiquing a Goliath industry and calling for broader education and external monitoring. During the process of editing my manuscript, the three identities and perspectives I had previously been juggling finally

[45] M. Zoll, *Cracked Open: Liberty and the Pursuit of High Tech Babies* (Northampton, MA: Interlink Books, 2013).

[46] P. Russell, "IVF tied to higher ovarian cancer risk?"

[47] L. Rapaport, "IVF linked to birth defects and childhood leukemia," *Huffington Post* (February 4, 2016). Available at: https://www.huffingtonpost.com/entry/reproductive-technology-linked-to-birth-defects-childhood-leukemia_us_56b3729ae4b04f9b57d89661. Accessed on: February 20, 2018.

fused into one clear voice. Making sense of the data I had been wrangling with, and synthesizing the data into a cohesive analysis, helped me shed the persona of the naïve and wounded consumer into that of a confident advocate and public speaker.

The second shift was an intellectual and analytical paradigm shift. After years of time-consuming research, I stumbled on the following statement on the website of the European Society for Human Reproduction and Embryology (ESHRE): "Around 1.5 million ART cycles are performed each year worldwide, with an estimated 350,000 babies born."[48] When I stopped to think about these numbers I realized they meant that annually there were on average 1,150,000 million ART *failures* worldwide, which translated into close to an 80 percent failure rate. I registered this information in a strikingly different manner from the publicized success stories so widely celebrated by the industry (and the media). I remember mulling that 80 percent statistic over and over again in my mind until it became clear that labeling IVF as "successful" didn't make sense because every year the vast majority of couples that underwent IVF experienced devastating failures – not just me.

Then it finally dawned on me, the former-repro-tech-consumer-now-wearing-her-investigative-journalism-hat, that the notion of IVF as "successful" was an ever so useful industry-constructed framework. From a corporate public relations perspective it was, and remains, a clever and effective strategy. If you conceal the millions of women who do not become mothers from public view and spotlight only those who deliver (healthy) babies, those opting for infertility services will not know to ask about overall failure rates or the limitations and risks associated with the procedures. And if no one knows to ask, the industry is not obliged to tell.

This contextual scaffolding mirrors Orwellian doublespeak: Success rates or failure rates, the glass is three-quarters empty or one-quarter full. It's the same thing, says industry,[49] so why not focus on the successes and

[48] "ART fact sheet," ESHRE (European Society of Human Reproduction and Embryology, July 2014). Available at: https://web.archive.org/web/20160314000811/https://www.eshre.eu/guidelines-and-legal/art-fact-sheet.aspx. Accessed on: March 17, 2016. This fact sheet has been updated to "ART fact sheet – Press information," ESHRE (undated). Available at: http://new.eshre.eu/~/media/sitecore-files/Annual-meeting/Lisbon/ART-fact-sheet.pdf?la=en. Accessed on: February 20, 2018.

[49] S. Herships, "The rising cost of infertilty," *NPR Marketplace* (September 13, 2013). Available at: www.marketplace.org/2013/09/13/business/rising-cost-infertility. Accessed on: February 20, 2018.

give people a bit of hope to hang onto? From my consumer perspective, the industry knows full well that it is not the same thing, especially for despairing women and couples who cannot help but interpret so-called success rates (no matter how low they are) as a universe full of hope.

Fortunately, a number of other former ART consumers have written books about their experiences, and several highlight the mental health risks associated with too much emphasis being placed on hope.[50] The most recent examples are Australian author Julia Leigh's 2016 book, *Avalanche: A Love Story*, and Jessica Hepburn's 2013 memoir, *The Pursuit of Motherhood*. And as John Brown, the father of the world's first IVF baby, said so adeptly in an interview a number of years ago: "We were desperate – [IVF] was one glimmer of hope so we grasped it."[51]

As I sat at the kitchen table making final edits to my manuscript, I thanked the ESHRE for my down-to-the-wire "Aha" moment. With the European repro tech industry and the International Committee Monitoring Assisted Reproductive Technologies (ICMART)[52] acknowledging an 80 percent per IVF cycle failure rate, I decided to make that the primary message of my book and publicity tour. I figured that if the public and the media were encouraged to view IVF from the perspective of "failures" rather than "successes," then perhaps there would be more rigorous inquiry and investigation into the industry's marketing and other questionable practices.

To this end, I coauthored an editorial with Pamela Tsigdinos, another repro tech consumer/patient whose book and blog, *Silent Sorority*,[53] serves as a lifeline to couples around the world who have withstood failed ART. Our op-ed, "Selling the Fantasy of Fertility," ran in the

[50] M. Zoll, *Cracked Open*. P.M. Tsigdinos, *Silent Sorority: A (Barren) Woman Gets Busy, Angry, Lost and Found* (BookSurge Publishing, 2009). "Book Intro" available at: http://silentsorority.com/book introduction/. Accessed on: February 20, 2018. J. Leigh, *Avalanche: A Love Story* (New York, NY: W.W. Norton, 2016). J. Hepburn, *The Pursuit of Motherhood* (Leicestershire: Matador, December 2013).

[51] S. Brown, *ESHRE: The First 21 Years* (Oxford: Oxford University Press, 2005), p. 7. Available at: www.eshre.eu/~/media/sitecore-files/About-ESHRE/The-first-21-years.pdf?la=en. Accessed on: February 20, 2018.

[52] "The world's number of IVF and ICSI babies has now reached a calculated total of 5 million," ESHRE (European Society of Human Reproduction and Embryology, July 2, 2012). Available at: https://www.sciencedaily.com/releases/2012/07/120702134746.htm. Accessed on: February 20, 2018.

[53] Silent Sorority (website, 2018). Available at: http://blog.silentsorority.com. Accessed on: February 20, 2018.

New York Times on September 11, 2013.[54] It immediately went viral on the Internet, making it one of the paper's most shared articles of that day[55] and giving the industry a run for its money.[56] Our rare consumer perspective contesting the dominant narrative of success and happiness, combined with the ESHRE and ICMART data confirming an 80 percent IVF failure rate per cycle, drew international media attention to the industry's aggressive marketing tactics. According to Muckrack, a journalism website that monitors and tracks the news, roughly 3,000 media outlets and 234,000 journalists[57] from New Jersey to Nigeria[58] reprinted, shared, or talked about our op-ed during news cycles. In addition, infertility and cancer support group chat rooms[59] discussed and debated the content of the op-ed for several weeks.

[54] M. Zoll and P. Tsigdinos. "Selling the fantasy of fertility."

[55] "*The New York Times* op-ed: 'Selling the fantasy of fertility,'" Silent Sorority (undated). Available at: http://silentsorority.com/new-york-times-op-ed-selling-fantasy-fertility/. Accessed on: February 20, 2018.

[56] "Who shared my link?", *Muck Rack*. Available at: https://muckrack.com/whoshared/?. Enter the following URL in the provided "Calculate shares" search box: http://www.nytimes.com/2013/09/12/opinion/selling-the-fantasy-of-fertility.html. Access date: February 21, 2018.

[57] "Who shared my link?"

[58] "Op-ed contributors: 'Selling the fantasy of fertility,'" *International Herald Tribune/ Nigeria Sun* (September 12, 2013). Available at: www.nigeriasun.com/index.php/sid/217014896/scat/45d771c7290844e9. Accessed on: October 27, 2016. No longer publicly available; on file with author. "Op-ed: 'Selling the fantasy of fertility,'" Medical news from *New York Times – Health* as posted by *Medical Web Times* (September 11, 2013). Available at: www.medicalwebtimes.com/read/oped_contributors_selling_the_fantasy_of_fertility/. Accessed on: October 25, 2016. No longer publicly available; on file with author. "Viewpoints: GOP 'obsession' with health law prevents progress on key financial issues; The 'fantasy of fertility'; VA is 'model' for transforming U.S. medicine," *KHN (Kaiser Health News) Morning Brief*: Summaries of health policy coverage from major news organizations (September 12, 2013). Available at: https://khn.org/morning-breakout/opinions-150/. Accessed on: February 20, 2018.

[59] Comments on: "Selling the fantasy of fertility," Life Without Baby (September 12, 2013). Available at: http://lifewithoutbaby.com/2013/09/12/selling-fantasy-fertility/. Accessed on: February 20, 2018. G. Quinn, "Response to 'Selling the fantasy of fertility? *NYT* 9/11/13 Op-ed,'" *Livestrong.org* (September 27, 2013). Available at: http://blog.livestrong.org/2013/09/27/response-to-selling-the-fantasy-of-fertility-nyt-91113-op-ed/. Accessed on: October 27, 2016. No longer publicly available; on file with author. T. Davidson, "The responsibility of a *New York Times* op-ed piece," *Davidson Communications* (undated). Available at: http://terridavidsoncommunications.com/the-responsibility-of-a-new-york-times-op-ed-piece-2/. Accessed on: February 20, 2018. H. Adrienne, Response to: "Selling the fantasy of fertility!" *Mind-Body-Unity* (October 2013). Available at: http://mind-body-unity.com/response-to-selling-the-fantasy-of-fertility/. Accessed on: October 25, 2016. No longer publicly available; on file with author.

During subsequent media interviews,[60] I described how the ART industry has dictated the discourse around repro tech for almost forty years[61] and that, out of self-protection, it omits information that could damage its reputation.[62] For example, poor record-keeping in many countries makes it virtually impossible to know exactly how many IVF babies have been born worldwide since 1978. Despite this, industry representatives estimate[63] and the media then reports that number to be between four and five million.[64] If we accept this estimate as more or less accurate, and if we then apply the ESHRE/ICMART statistic of an 80 percent failure rate over the entire four decades, it is very likely that there have been more than *20 million* failures.[65] And this would likely be an underestimate, as the failure rates would have been higher than 80 percent in the early years.

With my book's publicity campaign in full swing, I received hundreds of emails from people all over the world who shared their stories with me. A couple from Bogota, Columbia, told me they had tried IVF multiple times. Their last IVF cycle had resulted in the premature birth of twins that died at six weeks. When the grief-stricken couple returned to the clinic to speak to the doctor, they found the doors closed, the office abandoned. A woman from California emailed me that she had endured six failed IVF cycles, including three miscarriages, but the seventh cycle

60 A. Klein, "Give I.V.F. a chance," *New York Times – Motherlode* (September 17, 2013). Available at: http://parenting.blogs.nytimes.com/2013/09/17/give-i-v-f-a-chance/. Accessed on: February 20, 2018. S. Herships, "The rising cost of infertility." K. Brown, "Can fertility clinics deliver on their promises?," 90.9 WBUR – NPR News Station, Boston (August 12, 2013). Available at: http://legacy.wbur.org/2013/08/12/fertility-indus try-deliver. Accessed on: February 20, 2018. M. Zoll appears in the following: "The dark side of the infertility industry," the Leonard Lopate Show, National Public Radio (July 23, 2013). Available at: www.wnyc.org/people/miriam-zoll/. Accessed on: February 20, 2018. C. Adler, "No, 40 is NOT the New 25," *The Charles Adler Show* (Canada, undated). Available at: https://soundcloud.com/thecharlesadlershow/no-40-is-not-the-new-25. Accessed on: February 20, 2018.

61 M. Zoll, "IVF at 35: Science still hasn't outsmarted mother nature," *I.M.H.O. – Medium* (April 29, 2013). Available at: https://medium.com/i-m-h-o/ivf-at-35-science-still-hasnt-outsmarted-mother-nature-1a7e53f9d0#.kzeh8n9wp. Accessed on: February 20, 2018.

62 K. Elder and M.H. Johnson, M., "*The Oldham Notebooks* II. The treatment cycles and their outcomes."

63 B. Rochman, "5 million babies born through IVF."

64 B. Rochman, "5 million babies born through IVF." C. Russell, "Four million test-tube babies and counting," *The Atlantic* (October 7, 2010). Available at: https://www.theatlantic.com/technology/archive/2010/10/four-million-test-tube-babies-and-counting/64198/. Accessed on: February 20, 2018.

65 M. Zoll, "Questioning the cult of repro tech."

resulted in the birth of a healthy baby. While she considered herself a "success story," she was still grieving over her losses and applauded me for making public something so painful and taboo.

Some reproductive endocrinologists told me they agreed with my assessment of the industry, while others thought I was trampling on an important medical breakthrough that deserved more credit than I was assigning it. Doctors from neonatal intensive care units (NICUs) provided numerous examples of couples treated by IVF clinicians that later delivered premature twins or triplets.[66] Many of them had not been warned about the serious infant and maternal health risks – or fiscal consequences – associated with multiple embryo transfers, and had not been advised to purchase adequate health insurance. In some of these cases the families declared bankruptcy and the state had to intervene and pay for the costs of hospitalization and ongoing therapeutic care.[67]

Armed with this new supply of personal testimony and information, I authored dozens of articles that were published in health journals and online news sites, such as the University of Southern California's Health Journalism website, Reporting on Health,[68] Impact Ethics,[69] the Public Broadcasting System (PBS),[70] and RH Reality Check.[71] I also shared this stockpile of data with as many reporters and editors as I could so that they might weave nonindustry perspectives into their articles and programming about ART and IVF.

I spent hours monitoring daily news stories and following up with reporters who often only interviewed industry insiders. I respectfully suggested that they might want to balance future stories by also incorporating the perspectives of women or couples whose IVF outcomes were

[66] M. Zoll, "Hospital delivery charges significantly higher." M. Zoll, personal email correspondence with members of the National Perinatal Association (NPA) in the United States.

[67] M. Zoll, "Hospital delivery charges." M. Zoll, personal email correspondence with members of NPA.

[68] M. Zoll, "Stillbirths and infant health risks higher in California's artificially conceived infants," Center for Health Journalism, University of Southern California (February 28, 2014). Available at: http://bit.ly/2esSHMZ. Accessed on: February 20, 2018.

[69] M. Zoll, "ART ethics and health risks in the unregulated market," *Impact Ethics* (April 25, 2014). Available at: http://bit.ly/2ehulcr. Accessed on: February 20, 2018.

[70] M. Zoll, "What Apple and Facebook don't know about egg freezing," *PBS – To the Contrary* with Bonnie Erbe (October 15, 2014). Available at: http://to.pbs.org/1Cnignr. Accessed on: February 20, 2018.

[71] M. Zoll, "A note of caution: Freezing eggs is not a silver bullet for age-related infertility," *Rewire* (May 15, 2013). Available at: http://bit.ly/2dAopu7. Accessed on: February 20, 2018.

not successful, or inserting quotes from doctors outside the industry, like high-risk obstetricians, perinatologists, mental health specialists, and women's health advocates. Most of the time the reporters thanked me for providing them with an alternative viewpoint.

With the fortieth anniversary of IVF in 2018, I have relentlessly been conveying through interviews, articles, and essays that it is time to *thoughtfully* reflect on what has transpired in clinics and research laboratories since 1978. In public forums and on college campuses, I have asked the following sorts of provocative questions: What might happen if more patients, consumers, and parents of IVF babies began to speak out about their ART odysseys, and shared with the general public both positive and negative stories (and everything in between)? Could their amplified voices help expand the public dialogue about science and technology's role in our lives? Why not add to the discourse on the history of repro tech the testimonies of the 280 research subjects from the early experiments in England,[72] women who are now in their sixties and seventies who might want to share their recollections of Oldham Hospital or be recognized for their bodily sacrifices?

And what about the experiences of the estimated 20 million couples who have experienced failed IVF? I imagine that the inclusion of their narratives might shatter the dominant corporate narrative of "all is well." Such additions to the ART storyline might move the discussion beyond the superficial pro/con binary to foster a richer, more meaningfully balanced discussion that can usefully inform civil society's ongoing efforts to make informed policy decisions about these fast-developing technologies.

Since my book was released in 2013,[73] I have been told more than once that I should not be so disparaging of the repro tech industry because "people who can't have kids naturally need all the hope they can get." I can't tell you how many times friends and relatives of couples undergoing IVF have thanked me for writing my book and in the same breath have said something along the lines of: "I don't dare give it to my sister now. She would be so upset if she read it."

[72] K. Elder and M.H. Johnson, "*The Oldham Notebooks* II. The treatment cycles and their outcomes."

[73] M. Zoll, *Cracked Open.*

Those are moments when I force myself to remember what it felt like living from one IVF cycle to the next. I know from experience that most couples embedded in the world of ART are intensely focused on their conception journey and viscerally protective of any slivers of hope they can hold close to their hearts. Recognizing and respecting this, I decided to target my advocacy and education efforts to women and men between the ages of eighteen and thirty who might not yet be in the throes of thinking seriously about family planning or treatment for infertility.

When I am invited to speak at colleges, I candidly discuss the consequences of my own decisions to delay childbearing and naïvely enter into the realm of ART. The multimedia presentations I deliver are hard-hitting and include two powerful video clips. One features a woman injecting drugs into her abdomen and concludes with a tabletop covered with dozens of vials and syringes. The other video clip is a media montage of magazine covers and news headlines highlighting how IVF, sperm banks, egg selling and freezing, commercial surrogacy, and other ART-related topics are featured in the popular culture.

I also project photos of tiny premature ART babies struggling to survive on respirators in NICUs. I do so knowing full well that the images are unsettling. I want people to understand that due to the high number of multiple and premature births associated with ART, the CDC states that ART plays a substantial role in poor birth outcomes nationally.[74] I believe people should know that nearly 35 percent of ART conceived babies are twins, triplets, and higher order births, and that 31 percent are born premature.[75] As a result, these babies face higher rates of early death and those who do not die face serious health conditions, often with life-long implications.[76]

After I conclude my presentation, it is not unusual to have several students challenge the information that I have provided. Some disclose

[74] S. Sunderam, D.M. Kissin, S.B. Crawford, S.G. Folger, S.L. Boulet, L. Warner, and W.D. Barfield, "Assisted reproductive technology surveillance – United States, 2015," *Morbidity and Mortality Weekly Report*, Surveillance Summaries, Centers for Disease Control and Prevention, 67:3 (2018), 1–28. Available at: https://www.cdc.gov/mmwr/volumes/67/ss/ss6703a1.htm. Accessed on: February 19, 2018.

[75] S. Sunderam et al., "Assisted reproductive technology surveillance – United States, 2015."

[76] S. Sunderam et al., "Assisted reproductive technology surveillance – United States, 2015."

that they were born via IVF and that their health is fine. They want to know why I spend so much time talking about infant health risks. Others reveal that they have IVF cousins, nieces, nephews, and neighbors. They ask why I spend so much time discussing high failure rates. Some suggest that since I am from an older generation I couldn't possibly understand how liberating the option of egg freezing really is for young women today. When I explain that egg freezing is fraught with poor efficacy rates and potential health risks,[77] some accuse me of scaremongering.

In response, I say: "I believe you have the right to be informed about the full range of benefits, harms, and ethical concerns linked to repro tech. For most of your life, you've heard mostly about the positive outcomes. I think having a balanced and broader understanding of repro tech and its effect on your life and society at large is important and necessary." I let them know that a lot of the information I share with them required me, an average Internet user and ART consumer, many years to uncover. I invite them to think about how the existence of ART informs their perceptions about their bodies, their natural capacity to conceive and birth babies, and their future plans for career and family.

I know I frighten some young women and men with information they do not want to hear. I show them research findings that clash with the erroneous messaging they receive from the media, Twitter, and their friends' Facebook pages. I challenge their belief systems and their fantasies of being unchained from their reproductive destiny. I discuss the unavoidable natural decline of human fertility[78] in the middle of their highly productive third decade of life.

[77] P.M. Tsigdinos, "The sobering facts about egg freezing that nobody's talking about," *Wired* (October 2014). Available at: www.wired.com/2014/10/egg-freezing-risks/. Accessed on: February 20, 2018. J. Johnston and M. Zoll, "Is freezing your eggs dangerous?"

[78] D. Dunson, D.D. Baird, and B. Colombo, "Increased infertility with age in men and women," *Obstetrics and Gynecology*, 103:1 (January 2004), 51–56. B. Colombo and G. Masarotto, "Daily fecundability: First results from a new data base," *Demographic Research*, 3:5 (September 6, 2000), 1–39. Available at: www.demographic-research.org/Volumes/Vol3/5/. Accessed on: February 20, 2018. L. Wise, "Fertility peaks around age 30 for both males and females, Boston University study finds," Boston University School of Medicine (July 13, 2013). Available at: www.bumc.bu.edu/busm/2013/07/02/fertility-peaks-around-age-30-for-both-males-and-females-boston-university-study-finds/. Accessed on: February 20, 2018.

By talking about ART and IVF babies living in incubators for six months,[79] and those with cerebral palsy and stunted growth,[80] I contradict the dominant image of the happy-ending stories.

I passionately believe that to make informed decisions about their fertility and their futures, reproductive-aged women and men need to know about all aspects of the repro tech industry – the good, the bad, and the ugly. "I don't think women are so fragile that they can't handle the truth," I say gently but firmly. "I think it is better to plan your life, and if you so choose, plan your family, based on evidence and facts rather than marketing slogans and illusions."

There is usually silence in the auditorium after I say this. I know some in the audience are dismissive of what I have to say, but others are actually listening and processing the dissolution of the dual myths of science-can-fix-everything and women-can-have-it-all. These are the students standing right at the crossroad where the development of repro tech and the bright promises of the modern women's movement collide every day. Some may recognize that racial, class, and gender inequities will likely not be rectified by big industry paying poor women for their reproductive tissue,[81] or for their biological ability to gestate new life in exchange for money.[82] I hope they see beyond the seemingly benign

[79] T.A. Merritt, M. Goldstein, R. Philips, R. Peverini, J. Iwakoshi, A. Rodriquez, and B. Oshiro, "Impact of ART on pregnancies in California: An analysis of maternity outcomes and insights into the added burden of neonatal intensive care," *Journal of Perinatology*, 34:5 (May 2014), 345–250. M. Zoll, "Stillbirths and infant health risks."

[80] L.I. Kelley-Quon, C.H. Tseng, C. Janzen, and S.B. Shew, "Congenital malformations associated with assisted reproductive technologies: A California statewide analysis," *Journal of Pediatric Surgery*, 48:6 (June 2013), 1218–1224. J.L. Marino, V.M. Moore, K.J. Willson, A. Rumbold, M.J. Whitrow, L.C. Giles, and M.J. Davies, "Perinatal outcomes by mode of assisted conception and sub-fertility in an Australian data linkage cohort," *PLoS One*, 9:1 (January 8, 2014), e80398. Available at: http://journals.plos.org/plosone/article?id=10.1371/journal.pone.0080398. Accessed on: February 20, 2018. O. Lidegaard, A. Pinborg, and A.N. Andersen, "Imprinting diseases and IVF: Danish National IVF cohort study," *Human Reproduction*, 20:4 (April 2005), 950–954.

[81] D.E. Roberts, "Race, gender, and genetic technologies: A new reproductive dystopia?," *Signs*, 34:4 (Summer 2009), 783–804. M. Nahman, "Reverse traffic: Intersecting inequalities in human egg donation," *Reproductive Medicine Online*, 23:5 (November 2011), 626–633.

[82] S. Banu, "Colonial legacies, postcolonial biologies: Gender and the promises of biotechnology," *Asian Biotechnology & Development Review*, 17:1 (March 2015), 15–36. A. Pande, "Transnational commercial surrogacy in India: Gifts for global sisters?," *Reproductive Biomedicine Online*, 23:5 (November 2011), 618–625. Available at: http://www.rbmojournal.com/article/S1472-6483(11)00411-1/fulltext. Accessed on: February 20, 2018.

intentions of so-called cutting-edge employers like Apple and Facebook[83] offering female workers the chance to succeed professionally as long as they delay their safest natural childbearing window and rely instead on risky and experimental egg freezing services and, later, gamble with IVF.[84]

I point out that beneath the veneer of coopted feminist language the repro tech industry uses to exalt the ways in which egg freezing can "liberate" and "empower"[85] women and provide them with more "choices"[86] there lurks a highly unimpressive efficacy and safety record.[87] The ASRM reports that only 2–12 percent of thawed eggs lead to a live birth in women younger than age thirty-eight.[88] Along with its sister trade and lobby group, the Society for Assisted Reproductive Technology,[89] and the American College of Obstetricians and Gynecologists,[90] these organizations only endorse egg freezing for medical reasons. They oppose elective egg freezing among healthy young women in large part

83 M. Tran, "Apple and Facebook offer to freeze eggs for female employees," *The Guardian* (October 15, 2014). Available at: http://bit.ly/2dLDCWO. Accessed on: February 20, 2018. C. Farr, "Apple, Facebook will pay for female employees to freeze their eggs," *Reuters* (October 14, 2014). Available at: http://reut.rs/2ehCJ9i. Accessed on: February 20, 2018.

84 F. Baylis, "Left out in the cold: Arguments against non-medical oocyte cryopreservation," *Journal of Obstetrics and Gynaecology Canada*, 37:1 (2015), 64–67. F. Baylis, "Left out in the cold: Seven reasons not to freeze your eggs." *Impact Ethics*, (October 16, 2014) Available at: https://impactethics.ca/2014/10/16/left-out-in-the-cold-seven-reasons-not-to-freeze-your-eggs/. Accessed on: February 19, 2018.

85 A. Cattapan, K. Hammond, J. Haw, and L.A. Tarasoff, "Breaking the ice: Young feminist scholars of reproductive politics reflect on egg freezing," *International Journal of Feminist Approaches to Bioethics*, 7:2 (Fall 2014), 236–247.

86 H. Mertes, "Does company-sponsored egg freezing promote or confine women's reproductive autonomy?," *Journal of Assisted Reproduction and Genetics*, 32:8 (August 2015), 1205–1209.

87 J. Johnston and M. Zoll, "Is freezing your eggs dangerous?" A.P. Cil, H. Bang, and K. Oktay, "Age-specific probability of live birth with oocyte cryopreservation: An individual patient data meta-analysis," *Fertility Sterility*, 100:2 (August 2013), 492–499. A. Motluk, "Growth of egg freezing blurs 'experimental' label," *Nature*, 476 (August 23, 2011), 382–383.

88 "Can I freeze my eggs to use later if I'm not sick?," Fact sheet from *ReproductiveFacts.org*, American Society for Reproductive Medicine (2014). Available at: http://bit.ly/2olygrT. Accessed on: February 20, 2018.

89 The Practice Committee of the American Society for Reproductive Medicine and the Society for Assisted Reproductive Technology, "Mature oocyte cryopreservation: A guideline," *Fertility and Sterility*, 99:1, 37–43.

90 "ACOG: Committee opinion No. 584: Oocyte cryopreservation," *Obstetrics & Gynecology*, 123:1 (January 2014), 221–222.

because there is little medical evidence that tells us it is safe for women or offspring or that it is effective for delaying childbearing.[91]

Yet despite these medical cautions and concerns, most infertility clinics and egg freezing companies in the United States[92] (often in partnership with pharmaceutical giants like EMD Serono) actively promote this still unproven elective service on their websites[93] or at fancy cocktail parties in major metropolitan areas.[94] After interviewing several women that attended egg freezing parties in Manhattan and Los Angeles,[95] I learned that in at least four instances, presenters were not sharing data about high failure rates linked to egg freezing. Nor were they informing attendees that if they froze their eggs today they would need to undergo IVF in the future. No one discussed the potential harms the freezing process may cause the cellular structure of eggs or subsequent embryos,[96] if any even resulted.

This egregiously irresponsible behavior, and the industry's blatant disregard for patient and consumer informed consent, infuriates me. Such fury often leads me to put pen to paper, and in this case, it spurred me to write a series of articles challenging the industry's position. The articles were published in the *New York Times*,[97] *The New Republic*,[98]

[91] The Practice Committee of the American Society for Reproductive Medicine and the Society for Assisted Reproductive Technology, "Mature oocyte cryopreservation." "ACOG: Committee Opinion No. 584."

[92] B. Rudick, N. Opper, R. Paulson, K. Bendikson, and K. Chung, "The status of oocyte cryopreservation in the United States," *Fertility and Sterility*, 94:7 (December 2010), 2642–2646.

[93] "The decision to freeze your eggs," Shady Grove Fertility (2016). Available at: www.shadygrovefertility.com/treatments-success/egg-freezing. Accessed on: February 20, 2018. "Egg freezing/fertility preservation," Boston IVF (2016). Available at: https://www.bostonivf.com/treatments/egg-freezing/. Accessed on: February 20, 2018.

[94] R.M. Henig, "Should you freeze your eggs? An egg freezing party is not a great place to find answers to this or other questions," *Slate* (September 30, 2014). Available at: http://slate.me/1v58Qu3. Accessed on: February 20, 2018. "EggBanxx egg freezing parties," EggBanxx (2016). Available at: www.eggbanxx.com/events. Accessed on: October 27, 2016. No longer publicly available; on file with author.

[95] "To freeze or not to freeze," Sunday Night, Australian TV (September 6, 2015). Available at: http://yhoo.it/2dxRS7X. Accessed on: February 20, 2018. M. MacVean, "Party for future moms, FYOE (Freeze Your Own Eggs)," *Los Angeles Times* (May 8, 2015). Available at: http://lat.ms/1Qthyh7. Accessed on: February 20, 2018.

[96] A. Motluk, "Growth of egg freezing."

[97] M. Zoll, "Freezing eggs puts women and infants' health at stake," *New York Times* (October 16, 2014). Available at: http://nyti.ms/2GxrHtP. Accessed on: February 20, 2018.

[98] J. Johnston and M. Zoll, "Is freezing your eggs dangerous?"

and the website of the Public Broadcasting System,[99] among other venues. I take heart knowing that they will exist in cyberspace for many years where they will hopefully provide the public with a nonindustry perspective on current egg freezing techniques.

When it comes to repro tech, there is no shortage of topics to write about, but it is a time-consuming, marathon effort. For every article or interview generated, dozens more are being promoted by individual clinics that have on-site salaried public relations and marketing experts.[100]

One of the greatest barriers to expanding public education about ART and to investigating questionable practices in the United States is the very limited or near absence of funding from outside the infertility or pharmaceutical industries. There is abundant fiscal support from clinics, drug companies, and even radio stations[101] that enable some couples to access ART,[102] but very few, if any, that support investigative reporting or consumer/patient and media education. Without it, the underbelly of the industry remains hidden.

Though more may exist, I am only aware of two foundations in the United States that have funded important research concerning women's and infant's health and human rights within the context of the global ART industry. In 2014 the John D. and Catherine T. MacArthur Foundation awarded a $200,000 multiyear grant to the Center for Genetics and Society (CGS) and Our Bodies Ourselves to investigate international cross-border commercial surrogacy and commercial egg selling.[103] The California-based Tides Foundation also supports CGS's general

[99] M. Zoll, "What Apple and Facebook don't know."

[100] G. Jones, "The BIGGEST marketing challenge facing fertility centers in 2016," Fertility Bridge (February 8, 2016). Available at: http://www.fertilitybridge.com/blog/thebiggest marketingchallengefacingfertilityclinicsin2016. Accessed on: February 20, 2018. "10 marketing tips for IVF fertility practices to make patients love you," Kaufer DMC (2014). Available at: https://kauferdmc.com/marketing-resources/10-marketing-tips-for-fertility-practices-to-make-patients-love-you. Accessed on: February 20, 2018.

[101] D. Bates, "Win a baby! Radio station condemned for competition giving away free IVF treatments," *Daily Mail* (September 8, 2011). Available at: http://dailym.ai/2eC1D7L. Accessed on: February 20, 2018.

[102] "Two Louisiana couples given chance to build their families through Fertility Answers' Gift of Hope IVF grant," *Fertility Answers* (May 23, 2016). Available at: www.fertilityanswers.com/two-couples-win-chance-start-family-ivf-grant/. Accessed on: February 20, 2018. J. Ludden, "Making parenthood a reality through IVF grants," *National Public Radio* (May 11, 2011). Available at: www.npr.org/2011/05/11/135358223/making-parenthood-a-reality-through-ivf-grants. Accessed on: February 20, 2018.

[103] J. Norsigian and M. Darnovsky, "Center for Genetics and Society and Our Bodies Ourselves receive MacArthur Foundation award," *Our Bodies Ourselves* (June 11,

public education mission to alert and engage the public about the risks to social justice and the public interest posed by particular applications of ART and human genetic technologies.[104]

More than forty years ago, the Ford and Rockefeller foundations provided multiyear cycles of support for the early IVF experiments that led to the first human IVF birth in Britain.[105] It seems only fitting that today these influential institutions known for supporting global health and human rights initiatives should invest in research to track the health of women who have undergone (or are undergoing) ART and the health of the babies born of ART. It would also be appropriate for these foundations to launch investigations into human rights abuses linked to the burgeoning gestational surrogacy market,[106] the sale of human eggs,[107] and the trafficking of IVF infants on the black market.[108]

2014). Available at: www.ourbodiesourselves.org/2014/06/center-genetics-society-bodies-receive-macarthur-foundation-award/. Accessed on: February 20, 2018.

104 "Our Partners | Partner Directory: 'Center for Genetics and Society',": 'Center for Genetics and Society,'" The Tides Foundation (2016). Available at: https://www.tides.org/impact-partners/explore-our-partners/. Accessed on: February 20, 2018.

105 M. Johnson, "Robert Edwards: The path to IVF," *Reproductive Bio Medicine Online*, 23:2 (August 2011), 245–262. M.H. Johnson, S.B. Franklin, M. Cottingham, and N. Hopwood, "Why the Medical Research Council refused Robert Edwards and Patrick Steptoe support for research on human conception in 1971," *Human Reproduction*, 25:9, 2157–2174.

106 K. Robertson, "Booming demand, state protections attract commercial surrogate birthing," *The Sacramento Bee* (August 5, 2016). Available at: www.sacbee.com/news/local/health-and-medicine/article94020937.html. Accessed on: February 20, 2018. A. Pande, "Commercial surrogacy in India: Manufacturing a perfect mother-worker," *Signs*, 35:4 (Summer 2010), 969–992. G. Gorbea, "Mexico's booming business of producing babies for foreigners is about to go bust," *Vice News* (January 9, 2016). Available at: https://news.vice.com/article/mexicos-booming-business-of-producing-babies-for-foreigners-is-about-to-go-bust. Accessed on: February 20, 2018.

107 S. Carney, "The Cyprus scramble: An investigation into human egg markets," *Pulitzer Center on Crisis Reporting* (August 10, 2010). Available at: https://pulitzercenter.org/reporting/cyprus-scramble-investigation-human-egg-markets. Accessed on: February 20, 2018. J. Urist, "How much should a woman be paid for her eggs?," *The Atlantic* (November 4, 2015). Available at: www.theatlantic.com/health/archive/2015/11/how-much-should-a-woman-be-paid-for-her-eggs/414142/. Accessed on: February 20, 2018. A. Jones, "Putting a price on a human egg: Lawsuit claims price guidelines used by fertility clinics artificially suppress the amount women can get for their eggs," *Wall Street Journal* (July 26, 2015). Available at: www.wsj.com/articles/putting-a-price-on-a-human-egg-1437952456. Accessed on: February 20, 2018.

108 A. Ronan, "Inside the dark realities of the international surrogacy industry," *New York Magazine* (March 2015). Available at: http://nymag.com/thecut/2015/03/dark-side-of-international-surrogacy.html. Accessed on: February 20, 2018. G. Toboni, "Outsourcing embryos," *VICE/HBO* (October 2015). Available at: www.hbo.com/vice/episodes/03/

Their power and money helped spawn the industry and should now be used to expose resulting harmful and abusive practices.

In closing, I have no way of measuring the impact of the advocacy work I have dedicated myself to for the last several years. I can report, however, that through the synchronized efforts of members of the global watchdog network of which I am a member, more news outlets in the United States,[109] Australia,[110] and the United Kingdom and Europe[111] have begun to report about IVF in the context of failures and potentially risky health outcomes.

It is also worth noting that when Facebook and Apple announced that they would pay for their healthy female employees to electively freeze their eggs for nonmedical reasons, my extensive survey of global media found that the opinions of those opposing this policy were far stronger than those in favor of it.[112] Notably, a French newspaper included my perspective on the subject along with that of their irate Minister of Health, Marisol Touraine, who vehemently admonished employers for

26-lines-in-the-sand-and-outsourcing-embryos/synopsis.html. Accessed on: February 20, 2018. M. Goodwin, "Baby markets and the new motherhood: Reproducing hierarchy in commercial intimacy," *Huffington Post* (May 13, 2016). Available at: https://www.huffingtonpost.com/michele-goodwin/baby-markets-the-new-motherhood_b_7263050.html. Accessed on: February 20, 2018.

109 J. Brody, "The misleading promise of I.V.F." J. Brody, "Some I.V.F. experts discourage multiple births."

110 S. Dingle, "IVF doctors misleading women."

111 A. Magee, "Why are so many British women travelling abroad for fertility treatment?," *The Telegraph* (March 28, 2015). Available at: www.telegraph.co.uk/women/mother-tongue/11482483/Fertility-treatment-Why-British-women-are-travelling-abroad.html. Accessed on: February 20, 2018. Z. Williams, "Where's all that grief going? There may be five million IVF success stories, but for many millions more women, the treatments have failed. So why do we never hear from them?," *The Guardian* (September 27, 2013). Available at: www.theguardian.com/society/2013/sep/27/ivf-where-all-grief-going. Accessed on: February 20, 2018.

112 C.C. Miller, "Freezing eggs as part of employee benefits: Some see darker message," *New York Times* (October 14, 2014). Available at: www.nytimes.com/2014/10/15/upshot/egg-freezing-as-a-work-benefit-some-women-see-darker-message.html?_r=0. Accessed on: February 20, 2018. H. Minter, "By offering to freeze their employees' eggs, Apple and Facebook make it clear they don't know what women want," *The Guardian* (October 15, 2014). Available at: https://www.theguardian.com/women-in-leadership/2014/oct/15/apple-facebook-egg-freezing-employee-perk. Accessed on: February 20, 2018. M. Goodwin, "How Facebook and Apple's egg-freezing policies hurt working women," *AlterNet* (October 18, 2014). Available at: www.alternet.org/how-facebook-and-apples-egg-freezing-policies-hurt-working-women. Accessed on: February 20, 2018.

trying to dictate the reproductive trajectory of their female workers.[113] Alongside mainstream media's coverage of the new egg freezing "benefits," commentaries in the academic literature lamented the fact that businesses were more willing to pay for women to undergo experimental medical procedures than to establish policies that would support equal pay for equal work, extensive maternity and paternity benefits, flexible hours, or on-site workplace daycare.[114]

The fortieth anniversary of IVF is July 25, 2018.[115] In anticipation of the anniversary, several former ART consumers, myself included, have created a Twitter campaign using the hashtag #ReproTechTruths.[116] Our goal is to create an open space where women and men who endured failed IVF can post their testimony, along with those who may also have experienced health problems linked to ART – including offspring.

Whether they were research subjects,[117] experienced successful or failed IVF, bore stillborn infants or children with health challenges, sold or froze their eggs, or served as commercial gestational surrogates, those who feel comfortable enough to speak out can provide an important public service during the upcoming anniversary year and beyond. Personal testimony about repro tech is a powerful tool that can help individuals and civil society make decisions about repro tech and help direct public policy in the future. If you have a story tell, I hope you join us.

[113] M. Touraine, "La congélation des ovocytes 'n'est pas un débat pour directeurs de ressources humaines,'" *Gènéthique* (October 20, 2014). Available at: www.genethique.org/fr/marisol-touraine-la-congelation-des-ovocytes-nest-pas-un-debat-pour-directeurs-de-ressources-62292#.VEZwd4vF-EB. Accessed on: February 20, 2018.

[114] See, for example, F. Baylis, "Left out in the cold." H. Mertes, "Does company-sponsored egg freezing promote or confine women's reproductive autonomy?" M. Zoll, H. Mertes, and J. Gupta, "Corporate giants provide fertility benefits: Have they got it wrong?," *European Journal for Obstetrics, Gynecology, and Reproductive Biology*, 195 (December 2015), A1–A2.

[115] "The IVF global experience," Australian Research Council Discovery Project. "The world's first test-tube baby," *American Experience*, PBS.

[116] P.M. Tsigdinos, M. Zoll, et al., "#ReproTechTruths," Twitter correspondence (October 17, 2016). Available at: https://twitter.com/search?q=%2320MFailedIVF&src=typd. Accessed on: February 21, 2018.

[117] W. Wrigley, "IVF 'guinea pig' nearly died: New book challenges ethics of research pioneers," *Gisborne Herald* (February 20, 2018). Available at: http://gisborneherald.co.nz/localnews/2300436-135/ivf-guinea-pig-nearly-died. Accessed on: February 21, 2018. S. Crashley, *Testimony of an IVF Guinea Pig* (Gisborne, New Zealand: Loveley Books, 2014).

7

Establishing Pediatric Palliative Care

Overcoming Barriers

JOEL E. FRADER

Long ago and somewhat far away, I cared for a four-year-old boy with leukemia. Back then, only a very small number of children survived cancer. As expected, months into his illness his blood counts showed relapse. He and his mother were back in the hospital for another attempt to induce a remission. Things were not going well. One afternoon I found myself sitting on the boy's bed talking with him about the need to start another intravenous line; his mother went off to get something in the cafeteria. Unexpectedly, he burst into tears. I expressed surprise, remarking that he had braved that sort of invasion without crying many times before. He looked at me sadly and said, more or less, "It's not about the IV, I'm scared of dying."

As the inexperienced intern caring for the child, I had absolutely no idea how to respond. I didn't think he had any idea what death entailed. After a few moments searching for the right thing to say, I replied, "Do you know what that means?" Over the next little while, through the sobs, he managed to tell me that he thought he would go to sleep and then go to heaven. He felt frightened that he wouldn't see his mother any more (his father had left them shortly after the leukemia diagnosis) or feel her comforting embrace. He was scared, did not want to be alone, and did not know how to talk with his mother about any of this. He feared that such talk would make her too sad or angry. I tried to reassure the boy that we would not abandon him no matter what happened and we would work hard to help him communicate with his mother.

Later, I sought help from the program's child psychiatrist. He provided some references from pioneers in pediatric thanatology, such as Stanford

Friedman,[1] Morris Green,[2] John Spinetta,[3] and Gerald Koocher;[4] he told me the boy's feelings were not unusual. He said children notice how those caring for them talk and behave, often discovering what parents and staff attempt to hide from them. The children conclude that silence means the adults cannot tolerate talking about the situation and so they maintain the hush out of a desire not to upset their loved ones. Nothing I tried with my patient or his mother opened up communication. He died several weeks later. He had had much love and support from his mother, but I doubt he ever found the particular reassurance he craved.

Several more of my patients died during my residency. Almost none of the deaths followed open and honest conversations about the goals of care among clinicians, parents, or patients. One particularly troubling incident involved a hospitalized older teenager with widespread bone cancer. His treatment had resulted in the destruction of his immune system, making him vulnerable to infection. One night he developed fever and signs of overwhelming infection with life-threateningly low blood pressure. He told his parents he felt exhausted, knew his cancer could not be cured, and he felt ready to die. The parents and the patient talked this over, agreed to focus on keeping him comfortable, and then told us – the residents on call that evening – what had been decided. Their decision seemed completely reasonable and we supported it. We phoned the attending oncologist, who reacted with rage. He insisted we transfer the patient to the intensive care unit, start medicines to maintain vital signs, and do everything possible to keep the boy alive. We said we could not do that; the attending hung up the phone. A little while later, the oncology fellow appeared, arranged the transfer of the patient over the family's objection, and attempted to support the boy's blood pressure overnight. The boy died early in the morning, separated from his parents. It all seemed very wrong. The oncologist protested vigorously to our department chairman who, with characteristic gentleness, supported our decision, if not our overt defiance.

[1] S.B. Friedman, "Care of the family of the child with cancer," *Pediatrics*, 40:3 Suppl. (1967), 498–504.

[2] M. Green, "Care of the child with a long-term, life-threatening illness: Some principles of management," *Pediatrics*, 39:3 (1967), 441–445. M. Green, "Care of the dying child," *Pediatrics*, 40:3 Suppl (1967), 492–497.

[3] J.J. Spinetta and L.J. Maloney, "Death anxiety in the outpatient leukemic child," *Pediatrics*, 56:6 (1975), 1034–1037.

[4] G.P. Koocher, "Talking with children about death," *American Journal of Orthopsychiatry*, 44:3 (1974), 404–411.

Situations like this upset me repeatedly. With encouragement from a faculty member I pursued a fellowship that allowed me to study how and why clinicians and families make decisions about life and death matters in pediatrics. Courses in sociology and medical ethics led to insights, but not so much to practical skills for conducting conversations with patients and families about goals and preferences for end-of-life of care. In the 1980s, one could find some literature about adults and death/dying (e.g., Kübler-Ross's work[5]) and hospice care for adults, mostly from England (e.g., from Saunders[6] and Twycross[7]), but palliative and hospice care for children did not exist in the United States.

At that time, a handful of psychologists, psychiatrists, social workers, anthropologists, and nursing scholars had begun to study how children understand death, especially the deaths of their loved ones, and some had started to look at how dying children knew and thought about their own impending deaths (see work from Bluebond-Langner,[8] Friedman,[9] Green,[10] Koocher,[11] Sahler,[12] Schowalter et al.,[13] and Spinetta and Maloney[14]). Nevertheless, even as medicine and nursing began developing palliative and hospice care for adults in the United States, few seemed interested in palliative and hospice care for children. I spent a lot of time thinking about this. I worked in tertiary care pediatrics and I often saw children dying in the hospitals where I trained and worked. From what I could observe, nobody wanted to focus time or energy on improving the experiences of dying children and their loved ones.

My circumstances changed and I had to concentrate on other matters for some years. I remained interested in the care of dying children but had no clear path allowing me to turn curiosity and concern into action.

[5] E. Kübler-Ross, *On Death and Dying* (New York, NY: Macmillan, 1969).

[6] C.M.S. Saunders, "Terminal care," in D.J. Weatherall, J.G.G. Ledingham, and D.A. Warrell (eds.), *Oxford Textbook of Medicine* (Oxford: Oxford University Press, 1983), pp. 1–28.

[7] R.G. Twycross, "Hospice care: Redressing the balance in medicine," *Journal of the Royal Society of Medicine*, 73:7 (1980), 475–481.

[8] M. Bluebond-Langner, *The Private Worlds of Dying Children* (Princeton, NJ: Princeton University Press, 1978).

[9] S.B. Friedman, "Care of the family of the child with cancer."

[10] M. Green, "Care of the child with a long-term, life-threatening illness." M. Green, "Care of the dying child."

[11] G.P. Koocher, "Talking with children about death."

[12] O.J.Z. Sahler (ed.), *The Child and Death* (St. Louis, MO: Mosby, 1978).

[13] J.E. Schowalter, J.B. Ferholt, and N.M. Mann, "The adolescent patient's decision to die," *Pediatrics*, 51:1 (1973), 97–103.

[14] J.J. Spinetta and L.J. Maloney, "Death anxiety."

I then started working with a multidisciplinary group of colleagues at the University of Pittsburgh to develop a program in medical ethics. We came from many different backgrounds: law, sociology, philosophy, history of medicine, psychiatry, internal medicine, pediatrics, and more. In addition to starting an ethics teaching program in the medical school, several of us developed a clinical ethics consultation service. That service became fairly busy rather quickly; like others around the country, we found that a large fraction (around two-thirds) of the requests for our consultative help regarded end-of-life care. And, again similar to our colleagues elsewhere, many of the cases involved failed communications between clinicians and patients/families – often inadequate candor about the patient's prognosis, rather than substantive ethical conflicts about the "right" thing to do. In addition, it seemed clear that better symptom management, especially adequate pain relief, might well have prevented conflicts between families and clinicians from developing in the first place. The practitioners among us soon found ourselves asking if systematic efforts to improve clinicians' relationship and symptom control skills could substantially reduce the demand for ethics consultation. These experiences led more than a few of us in clinical ethics to begin morphing into palliative care doctors and teachers. However, this process seemed to develop much more slowly in pediatrics compared with adult medicine. Why?

A brief aside: Two "big picture" factors almost certainly contributed to the slow development of pediatric palliative and hospice care. First, fortunately only about 55,000 children die each year in the United States.[15] That constitutes a small fraction (less than 2 percent) of all yearly deaths in the United States. Moreover, over half of those deaths happen very quickly.[16] Examples include death within seventy-two hours of the birth of a very premature infant, the rapid demise of a newborn with severe congenital abnormalities, demise shortly after a major traumatic event, or admission to the hospital with an overwhelming infection.[17] No matter how one looks at it, the statistics make it clear that childhood deaths in which pediatric palliative care clinicians might develop strong relationships with patients and

[15] M.J. Field and R.E. Behrman (eds.), *When Children Die: Improving Palliative and End-of-Life Care for Children and Their Families* (Washington, DC: National Academies Press, 2003).

[16] M.J. Field and R.E. Behrman (eds.), *When Children Die.*

[17] M.J. Field and R.E. Behrman (eds.), *When Children Die.*

families and intervene effectively do not happen frequently. Crudely put, the demand is relatively low.

Second, following important biomedical advances in the mid-twentieth century, the death rate among children from two major causes of death, namely, cancer and prematurity, dropped dramatically.[18] Long-term (more than five years) survival in childhood cancer went from a rarity to an overall expectation, now in the range of 75–80 percent.[19] Premature infants who commonly died with respiratory failure from inadequate lung development when born more than six to eight weeks before full term began to survive and thrive with the introduction of appropriately sized machines to support their breathing and medications to help their lungs stay open and functional.[20] Many infants born 15–18 weeks early now survive.[21] As these astonishing achievements developed, along with other advances in pediatric therapeutics, clinicians and families found it harder and harder to accept that death might actually occur during childhood. Everyone wanted to believe that no child had to die, at least not in tertiary care children's specialty hospitals.

Who needs pediatric palliative care in the face of triumphant, even miraculous, progress? Some who were engaged in clinical trials in pediatrics decried any decisions to limit attempts at rescue, even when the burdens of continuing efforts seemed too much for particular children to bear. Those zealots saw loving decisions by parents or other clinicians who said "enough already" as standing in the way of the scientific project designed to overcome *all* childhood ills. From that perspective, pediatric palliative care only threatened the march of science.

This is how things were in the 1980s through to the 1990s. Over the course of just a single century there had been incredible drops in childhood age-specific death rates, achieved through prevention (sanitation, immunization, economic development) and through remarkably successful treatments for major disorders that had previously caused many childhood deaths.[22] Pediatric subspecialists such as pediatric hematologists/oncologists, neonatologists, and pediatric critical care

[18] M.J. Field and R.E. Behrman (eds.), *When Children Die.*

[19] E. Ward, C. DeSantis, A. Robbins, B. Kohler, and A. Jemal, "Childhood and adolescent cancer statistics, 2014," *CA: A Cancer Journal for Clinicians*, 64:2 (2014), 83–103.

[20] J. Lantos and W. Meadow, *Neonatal Bioethics: The Moral Challenges of Medical Innovation* (Baltimore, MD: Johns Hopkins University Press, 2006).

[21] M.J. Field and R.E. Behrman (eds.), *When Children Die.* E. Ward et al., "Childhood and adolescent cancer statistics." J. Lantos and W. Meadow, *Neonatal Bioethics.*

[22] M.J. Field and R.E. Behrman (eds.), *When Children Die.*

specialists (and the institutions in which they worked) portrayed these accomplishments in glowing terms, eschewing any association with death or disability.[23] Supporting the development of pediatric palliative care risked undermining that image and scaring away potential institutional donors who would expect evidence of cures as return on their philanthropic investment. Selling triumph over leukemia or brain tumors is one thing; pitching the need to alleviate the pain of disseminated cancer or enhancing the ability of parents to face their lost dreams as associated with the death or serious disability of their child is quite another.

Nevertheless, some insiders failed to accept this rose-colored view. Despite the very best efforts, some children could not be saved from whatever unfair misery the universe imposed on them. Ann Goldman, a courageous British pediatric oncologist practicing at one of the most prestigious children's hospitals in the world, London's Great Ormand Street Hospital for Children, began to see that despite her field's great progress, a proportion of children with cancer would die from their malignancies.[24] Dr. Goldman also noted that modern medicine could help make the quality of life before death substantially better for her patients and their families. Borrowing from progress in adult palliative care in Britain and appreciating the desires of families and those treating pediatric cancer to push on as long as possible, Dr. Goldman introduced the notion that pediatric palliative care should commence from the time doctors diagnosed and first treated a child's cancer.[25] The idea has two great advantages: (1) Those who focus much of their clinical effort on a child's comfort (physical, emotional, and spiritual) can likely provide better symptom management than those focusing on killing every last cancer cell and (2) for those patients whom oncologists cannot cure, early involvement of palliative care clinicians permits the development of trusting relationships between families and those best equipped to manage end-of-life care.[26]

Many of us in pediatric palliative care in the United States have come to believe in, and aspire to implementing, the Goldman model of

[23] J. Frader, E. Morgan, T. Levinson, J. Morrow, J.M. Saroyan, M.J. Gilmer, and B.S. Carter, "Barriers, education, and advocacy in palliative care," in B.S. Carter and M. Levetown (eds.), *Palliative Care for Infants, Children, and Adolescents: A Practical Handbook* (Baltimore, MD: Johns Hopkins University Press, 2004), pp. 44–66.

[24] A. Goldman, S. Beardsmore, and J. Hunt, "Palliative care for children with cancer: Home, hospital or hospice?," *Archives of Disease in Childhood*, 65:6 (1990), 641–643.

[25] A. Goldman, "Palliative care for children," *Palliative Medicine*, 9:3 (1995), 179–180.

[26] A. Goldman et al., "Palliative care for children with cancer."

involving palliative care clinicians from the time of diagnosis of a life-threatening childhood illness. Doing that has proved hard in our country. Some pediatric subspecialists "battling" disease hold scientistic views about the power of their clinical tools. Some also claim that the introduction of palliative care will undermine hope with devastating psychological consequences for parents, patients, and perhaps the extended family. Yet no empirical evidence supports this notion. Moreover, some studies show that most parents in the United States want open, honest appraisals of what their children face and strongly prefer at least a major, if not the determinant, role in decisions about their child's care.[27] Nevertheless, paternalistic notions of "protecting" parents from feeling "guilty" about participating in decisions to limit treatment do not fade readily.

From another perspective, even if disease-focused clinicians opposed the growth and development of pediatric palliative care, administrators might have wanted to endorse the development of palliative care programs because of the potential to improve care, increase family satisfaction, and perhaps reduce unnecessary spending on expensive interventions in the last days, weeks, or months of life. However, as pediatric palliative care struggled to establish itself in the United States, "quality" health care typically meant more interventions (not fewer), and more interventions almost always meant more money for the hospital – so, an improved bottom line. Weeks or months in the hospital, better yet, in the pediatric intensive care unit, translated into filled beds with steady "fee-for-service" income.

Few institutions assessed family satisfaction with the care they provided, and when administrators did seek such information, thanatophobia generally prevented surveys from reaching the hands or ears of parents whose children had died. Institutional Review Boards (IRBs), intended to protect research subjects, routinely prohibited investigations of family coping and assessment of care once a child had died, on the grounds that the studies would upset bereaved parents. This has been true even though available empirical evidence does not support this paternalistic view.[28] Administrators, public relations people, and board

27 K.N. Michelson, T.K. Koogler, K. Skipton, C. Sullivan, and J. Frader, "Parents' reactions to participating in interviews about end-of-life decision making," *Journal of Palliative Medicine*, 9:6 (2006), 1329–1338.

28 K.N. Michelson et al., "Parents' reactions." K.S. Hoehn and J.E. Frader, "Approaching parents for organ donation: Who and when?," *Pediatric Critical Care Medicine*, 9:2

members continued to want "the community" to view their institutions as places of cure and happiness, fearing mention of, much less attempts to raise money for, endeavors associated with death, loss, and grief.

Interestingly, things seemed altogether different in adult medicine. Major philanthropies (the Robert Wood Johnson and the Soros foundations, most prominently) invested heavily on improving care of the dying.[29] It seemed dying after four score and ten deserved attention, dying in childhood not so much. To be fair, the Project on Death in America funded a few small efforts that did help to stimulate interest in pediatric palliative care across the United States. Even when those programs began to report success in helping dying children and their families, interested clinicians at major children's hospitals struggled to find support to develop palliative care services.[30] Some institutions established programs with psychosocial support provided by nurses.[31] However, up through the middle of the 1990s, very few tertiary care pediatrics programs in the United States had physician-led or even physician-involved pediatric palliative care.[32] That contrasted sharply with developments in Britain[33] and Canada.[34]

My own attempts to establish a pediatric palliative program at a large children's hospital in the 1990s came up against all of the barriers mentioned above. Many subspecialists confidently asserted that there

(2008), 234–235. U. Kreicbergs, U. Valdimarsadottir, G. Teineck, and J.-I. Henter, "A population-based nationwide study of parents' perceptions of a questionnaire on their child's death due to cancer," *Lancet*, 364:9436 (2004), 787–789.

29 I. Byock, J.S. Twohig, M. Merriman, and K. Collins, "Promoting excellence in end-of-life care: A report on innovative models of palliative care," *Journal of Palliative Medicine*, 9:1 (2006), 137–146. F. Aulino and K. Foley, "The Project on Death in America," *Journal of the Royal Society of Medicine*, 94 (2001), 492–495.

30 R.M. Hays, J. Valentine, G. Haynes, J.R. Geyer, N. Villareale, B. McKinstry, et al., "The Seattle Pediatric Palliative Care Project: Effects on family satisfaction and health-related quality of life," *Journal of Palliative Medicine*, 9:3 (2006), 716–728.

31 K.A. Mastro, J.E. Johnson, N. McElvery, and C. Pruester, "The benefits of a nurse-driven, patient- and family-centered pediatric palliative care program," *Journal of Nursing Administration*, 45:9 (2015), 423–428.

32 C. Feudtner, J. Wormer, R. Augustin, S. Remke, J. Wolfe, S. Friebert, and D. Weissman, "Pediatric palliative care programs in children's hospitals: A cross-sectional national survey," *Pediatrics*, 132:6 (2013), 1–8.

33 A. Goldman et al., "Palliative care for children with cancer."

34 K. Widger, D. Davies, D.J. Drouin, L. Beaune, L. Daoust, R.P. Farran, et al., "Pediatric patients receiving palliative care in Canada: Results of a multicenter review," *Archives of Pediatrics & Adolescent Medicine*, 161:6 (2007), 597–602. H. Siden, N. Chavoshi, B. Harvey, A. Parker, and T. Miller, "Characteristics of a pediatric hospice palliative care program over 15 years," *Pediatrics*, 134:3 (2014), e765–e772.

was no need for a program dedicated to either symptom management or psychosocial support for children with life-threatening conditions. Others insisted their clinical program already provided adequate care to dying children and their families. These claims persisted despite family feedback suggesting that end-of-life care had not adequately addressed pain or other troubling problems, and that previously involved clinicians often spent much less time with the patient or family after life-prolonging efforts ceased, leaving parents feeling abandoned. As well, residents in pediatrics reported that faculty did not (and perhaps could not) teach trainees about good end-of-life care. When hospital leaders finally approved an attempt to develop a pediatric palliative care program, they made it clear that clinicians interested in such a program would have to find independent (philanthropic) support for this new initiative and that they would have to do so without interfering with the institution's ongoing efforts to fund *life-saving* treatment and research.

I then moved to another department at Northwestern University in Chicago, Illinois, and with two other clinicians already informally involved in palliative care (each of whom had independently taken the time and effort to complete the palliative care certification examination), we established an institutionally sanctioned clinical palliative care program at Lurie Children's Hospital. This took almost a decade. And it would be almost another decade before the idea that the program deserved support as part of the hospital's regular operations budget (comparable to other clinical efforts) was accepted. In 2016 the Department of Pediatrics created a formal Division of Pediatric Palliative Care.

It is my view that in some parts of the United States one can find serious wariness of pediatric palliative care by clinicians and wholesale disconnections between them and families (and the communities from which they come) based on socioeconomic, religious, ethnic, or other demographic factors. Put starkly, sometimes when clinicians say they no longer believe they can affect a cure or even sustain life beyond a few days or weeks, some families say: "Well, it's not up to you, it's in God's hands so keep trying," and/or "We think you would not say that if our child were [rich/Christian/white]." Some families believe that largely white and relatively well-off clinicians harbor inevitable and deep biases against whichever group with whom they identify.[35]

[35] K.L. Moseley, A. Church, B. Hempel, H. Yuan, S.D. Goold, and G.L. Freed, "End-of-life choices for African-American and white infants in a neonatal intensive-care unit: A pilot study," *Journal of the National Medical Association*, 96:7 (2004), 933–937. A.K. Smith,

These families mistrust clinicians who attempt to justify recommendations based on statistical or other evidence.

In these cases, frustrated clinicians sometimes call on palliative care teams to get the Do Not Resuscitate (DNR) consent form signed or to convince the family to limit treatment in other ways, such as stopping mechanical ventilation or dialysis. This rarely succeeds. The ability of palliative care clinicians to help families share in making decisions about their severely ill children depends on trust. Only a few gifted practitioners can rapidly develop deep, empathic rapport with angry or terribly disappointed families, especially when others' therapeutic alliances have broken down or never been established in the first place. Overcoming powerful cultural and socioeconomic divides in the face of a family crisis and acute grief is a nearly impossible task and one that usually takes long and persistent efforts. Palliative care clinicians need to help their colleagues set reasonable expectations for the outcome of palliative care consultations in fraught circumstances.

Pediatric palliative care has finally taken off, although clearly in some places in the United States it has climbed higher than others.[36] Many obstacles remain. We have few training programs and the process for achieving "board certification" absurdly requires pediatricians to pass the same examination as colleagues in internal medicine, family medicine, anesthesiology, etc. More than 90 percent of that exam pertains to adult medicine. For the foreseeable future, we will not have enough pediatric palliative care clinicians. Pediatric palliative care will never pay for itself; in Illinois we typically receive reimbursement at a rate of roughly 13 percent of what we bill. Those with attitudes that disparage pediatric palliative care still need to have their minds changed: Pediatric palliative care clinicians have knowledge, skills, and practices that can make a positive difference in the lives of patients and families, just as pediatric kidney or lung doctors have specific contributions to make. Convincing colleagues to use "automatic" triggers for palliative care consultation may help. In some intensive care units for adults, any patient who remains in the unit for more than a specified period (e.g., 7 or 10 days) has a

E.P. McCarthy, E. Paulk, T.A. Balboni, P.K. Maciejewski, S.D. Block, and H.G. Prigerson, "Racial and ethnic differences in advance care planning among patients with cancer: Impact of terminal illness acknowledgment, religiousness, and treatment preferences," *Journal of Clinical Oncology*, 26:25 (2008), 4131–4137. R. Rhodes and J.M. Teno, "What's race got to do with it?," *Journal of Clinical Oncology*, 27:33 (2009), 5496–5498.

36 C. Feudtner et al., "Pediatric palliative care programs in children's hospitals."

palliative care consultation. In pediatrics, perhaps we should routinely invoke palliative care consultation for any child with several kinds of cancer that carry grave prognoses (some brain tumors, other malignancies that are widely disseminated through the body at the time of diagnosis) or for children with severe, chronic, congestive heart disease, whether or not they become candidates for heart transplantation.

From an ethics perspective, offering pediatric palliative care enhances options for patients and families.[37] Those for whom we care deserve to know that under many circumstances we can provide expert help to alleviate symptoms, assist in reducing the psychological and spiritual suffering patients and family members experience, and support loving decisions, including some that involve limiting or forgoing treatment. Colleagues in various pediatric subspecialties need to learn that the efforts of pediatric palliative care clinicians do not undermine or threaten their expertise. Unlike what seems to be the case in the care of the elderly, increased use of pediatric palliative care probably will not save a great deal of money by bringing about dramatic reductions in unwanted hospitalizations, intensive care unit stays, or emergency department visits toward the end of life. However, substantial improvements in the quality of life of dying children and their family members may have many long-term dividends for those families and their communities.

[37] C. Feudtner and P.G. Nathanson, "Pediatric palliative care and pediatric medical ethics: Opportunities and challenges," *Pediatrics*, 133:Suppl. 1(2014), S1–S7.

8

History and Philosophy of Science Engaging the Public

JANE MAIENSCHEIN

The premise for this book is that academic study of ethics too often remains separate from ethics in action and that a growing number of scholars want to take their ethics actively into the real world. An underlying assumption is that action that serves as advocacy does not always find a comfortable home in the academic world. Other chapters in this volume present examples in which scholars have felt the need to step outside the academic world to promote and advocate for justice.

At Arizona State University (ASU), we are working toward a different model called the "New American University."[1] Here the university is very much explicitly and reflectively engaged with the world, and faculty members are encouraged to pursue "use-inspired" as well as "curiosity-driven" research. Those who embrace this approach have no problem pointing to important "broader impacts" in their proposals for federal grants, and they have developed a diverse set of education and training programs based on the values and intended goals of the new model. Along the way, assessment of each faculty member's teaching and research is supposed to consider impact and outcomes along with the usual measures of success, and "service" is to be taken not as a separate category, but rather as fully integrated.

This chapter starts by outlining this model for a New American University, taking a look at a few of the design principles on which it is built. Then it moves to a discussion of my own choices about how to become engaged in the community outside the academy and what it has meant to advocate for reasoned decision-making. My approach leads to a focus on the ways in which history and philosophy of science joins bioethics to inform understanding and action in the world today. Finally, I discuss the implications of this approach for connecting

[1] "New American University," Arizona State University (2016). Available at: https://newamericanuniversity.asu.edu/. Accessed on: October 28, 2016.

research and education through the example of the Embryo Project, which goes beyond what any one of us can do alone by drawing on collective action.

The ASU Model for a New American University

When I arrived at ASU in 1981, I initially pursued the traditional academic career path, writing books and articles, giving professional lectures, and carrying out academic service at ASU and for professional organizations. I gave occasional talks in the community and spoke to newspaper reporters about why evolution is important, but it was clear that those activities counted as "extra" work and were not particularly valued by my colleagues. This was a typical academic pattern at the time.

As chair of the philosophy department from 1991 to 1996, I began to see other possibilities as then-President Lattie Coor urged the academic leaders, including department chairs, to think seriously about ways to improve education as a matter of the university's civic responsibility to the state of Arizona. The state was then one of the lowest in the country in most measures related to educating its citizens (including per capita spending on education, percentage of students finishing high school, percentage of students who finished high school going on to college, and teacher salaries), and the state legislature did little to change that sad fact. Coor's leadership pointed to ways that we could each work with underprepared students arriving in our classes to help them succeed rather than weeding them out. This was a first step and a foundation for what the next and current President Michael Crow points to as a policy of inclusion. As our new university charter puts it: "[W]e are measured not by whom we exclude, but rather by whom we include and how they succeed."[2]

It always takes a while for new ideas to take root. In 2002, when President Crow arrived, past-President Coor's ideas became entrenched in our university's mission. Crow arrived with an expansive vision, which he articulated energetically in his inaugural address. Since then, he has brilliantly expanded, extended, and implemented that vision.[3]

[2] ASU, *Arizona State University Charter* (Office of the President, 2015). Available at: https://president.asu.edu/about/asucharter. Accessed on: October 28, 2016.

[3] *Higher Ed Mavericks: How Four Visionaries Are Leading the Charge for Change*, thematic issue of *The Higher Education Workplace*, 2:2 (Fall 2010), 14–28.

Not everybody is a fan or accepts the "design principles" or changes involved in transforming the university; yet many do, and today ASU is a different place.[4] It is different from what it was when I began my career, and, in many ways, it is different from most other academic institutions, though many are adopting and adapting the ideas as well. The vision and energy behind our new university charter allowed me to take a different path – allowed me to diverge from traditional expectations and to call for new measures of success. It is important to understand the values underlying this transformation, since few institutions have provided such strong support for community engagement as ASU is doing.

Crow's recent book with historian William B. Dabars, *Designing the New American University*, outlines the vision, principles, and goals of the university and places that discussion in a larger context of educational history. We need, they argue, a new approach to higher education. The model in which a relatively few elite institutions select their favorites and proudly display statistics about how many they excluded has to change. There is strong evidence that a more egalitarian approach can help promote productive "socioeconomic engineering,"[5] which should speed up and improve much-needed social transformations.

Research and education should embrace more transdisciplinary thinking that is globally engaged while also locally invested. Rather than reforming every institution in the same way, we need a set of design principles to help guide the building of each individual, contextually responsive institution. This diversity will yield a stronger result than seeking uniform goals. Let us rethink the social contract for science and knowledge more generally, Crow and Dabars urge, and move toward a knowledge economy based on discovery, creativity, and innovation all oriented toward increasing social good. Let us ask why we want to study nature through science, for instance, and how we can best learn from a history of wise, pragmatic thinkers.

One could be rather cynical about what may look like just slogans, and it is fair to ask whether all this rhetoric is also real. The answer is "yes"; Crow and his New American University are committed to putting the

[4] J. Warner, "ASU is the 'New American University' – it's terrifying," *Inside Higher Ed* (January 25, 2015). Available at: www.insidehighered.com/blogs/just-visiting/asu-new-american-university-its-terrifying. Accessed on: October 28, 2016.

[5] M.M. Crow and W.B. Dabars, *Designing the New American University* (Baltimore, MD: Johns Hopkins University Press, 2015), p. 30.

scholarship and ideas into practice. He uses his presidency for advocacy and engagement in much the same way the examples in this volume are asking people to do. Crow is a leader for a new University Innovation Alliance,[6] which brings together eleven research universities that teach nearly half a million students each year. The goal is to support all students, including those from lower income brackets, to succeed in college and gain the skills to succeed thereafter. Success means helping to make society better, by focusing on useful and valued outcomes. Success also means supporting faculty members who carry their scholarship into the world.

Thomas Jefferson had it right, Crow and Dabars believe, when he tied universities to the goals of democracy: "If a nation expects to be ignorant and free, in a state of civilization, it expects what never was and never will be."[7] Moreover, "no other sure foundation can be devised for the preservation of freedom and happiness . . . [than] the diffusion of knowledge among the people."[8] James Madison was also on the right track: "Knowledge will forever govern ignorance: And a people who mean to be their own governours must arm themselves with the power which knowledge gives."[9] On this view, in short, a university should function as a sort of use-inspired effort to improve society.

That goal means that education should build strong foundations and build on traditional scholarship, but we should also ask: Why and for what purpose? It should bring discussion of the problems to be addressed into discussions of how to address them. There is no one prescription for a new university, because each needs to be locally contextualized. Yet the design principles carry over (see Table 8.1). Above all, Crow and Dabars include integrating scholarship with action. This kind of setting has made it possible for me to take up my own approach to action, arguing for reasoned decision-making in pursuit of a better society.

6 "Who we are," University Innovation Alliance (2016). Available at: www.theuia.org/#about. Accessed on: October 28, 2016.

7 T. Jefferson (1816), as cited by M.M. Crow and W.B. Dabars, *Designing the New American University*, p. 28.

8 T. Jefferson (1816), as cited by M.M. Crow and W.B. Dabars, *Designing the New American University*, p. 28.

9 J. Madison (1822), as cited by M.M. Crow and W.B. Dabars, *Designing the New American University*, p. 28.

Table 8.1 *Design Principles for a New American University**

	Principle	Explication
1	Leverage Our Place	ASU embraces its cultural, socioeconomic, and physical setting.
2	Transform Society	ASU catalyzes social change by being connected to social needs.
3	Value Entrepreneurship	ASU uses its knowledge and encourages innovation.
4	Conduct Use-Inspired Research	ASU research has purpose and impact.
5	Enable Student Success	ASU is committed to the success of each individual student.
6	Fuse Intellectual Disciplines	ASU creates knowledge by transcending academic disciplines.
7	Be Socially Embedded	ASU connects with communities through mutually beneficial partnerships.
8	Engage Globally	ASU engages with people and issues locally, nationally, and internationally.

* M.M. Crow and W.B. Dabars, *Designing the New American University*, p. 243.

Exploring the Larger World

I direct the Center for Biology and Society,[10] which resides within the School of Life Sciences at ASU and embeds ethicists, historians, and philosophers of science in a life sciences unit. The Center's mission is to promote understanding of the diverse ways that biology plays out in society and to examine the nature of the intersections and how they have changed over time. It is a kind of center that could not quite exist in other places, because the administration values different kinds of actions and does not measure success in only the traditional academic ways where the emphasis is on publications and, increasingly, on external research funding (and, in some institutions, on student teaching evaluations) with scant attention to public education and engagement. History and philosophy of science provides a foundation for our work, which involves community education activities of many different sorts, including

[10] Center for Biology and Society, Arizona State University (2016). Available at: https://cbs.asu.edu/. Accessed on: October 28, 2016.

participating in more traditional venues and in developing new approaches to ethics-in-action for the biosciences.

In my case, the focus remains on embryos, stem cells, and, more recently, gene editing. This work is guided by the conviction that reason should trump intuition in guiding political choices: Decisions related to the biosciences should at the least be consistent with the best available scientific knowledge, understood both "in-the-now" and "in-the-past." Reason and reflection are virtues.

History and Philosophy of Science Informing Science and Society

While the current ASU vision has been established by Crow building on the strong foundations set in place by Coor, it was during Coor's presidency that I had my first chances to diverge from the traditional academic path toward engagement of a type that has worked beautifully for me. From January 1997 through December 1998, I had the wonderful opportunity to learn about the larger world when I served as a Congressional Fellow and senior science advisor to Arizona Congressman Matt Salmon during the 105th congressional session. This occurred because Salmon came to ASU and said to Coor, in effect, "I'm on the House Science Committee, I have ASU in my district, and I need help being informed about science." The congressman wanted someone with a broad view who was willing to work with him and help provide reliable information. Coor pointed to someone at the intersections of history and philosophy of science and the biological and biomedical sciences, someone who could advocate for scientific understanding rather than representing any one part of the scientific research enterprise. Taking on the role of Congressional Fellow involves a different sort of advocacy – namely, advocacy for grounding decisions on the best available evidence and information, committed to avoiding political advocacy.

Of course, any knowledge can be contested. There are obviously "deniers" for many scientific claims. This is where history and philosophy of science becomes useful. It is not necessary to take sides among those yelling at each other right now. A historical perspective can help illuminate what is at issue and how different positions have developed over time. Philosophical inquiry can help articulate underlying assumptions that shape how partisans see available information.

During the 105th Congress, I brought a history and philosophy of science perspective to understanding embryos. The year 1997 introduced cloning, with Dolly the sheep, and 1998 brought human embryonic stem cell research. In both cases, public outcry led congressmen to draft legislation. Most of this draft legislation was very poorly conceived and based on faulty assumptions. One proposed bill, for example, aimed to outlaw "genetic copies" and was written in such a way that it would have outlawed naturally occurring identical twins. Other proposed bills would have criminalized women's decision-making regarding whether and how many embryos to transfer by labeling decisions about selective embryo transfer as homicide. In these and other instances, many of the claims and the underlying reasoning were quite wrong. Bringing a history and philosophy of science perspective to bear on these debates informed the congressmen about what really was at issue and what we already knew from science.[11] Fortunately, none of these flawed draft bills ever reached the floor.

Working with Congress was hugely educational, and I'll point to two lessons in particular. First, I learned that decision-making rarely relies on serious (scientific or other) research and typically draws heavily on party politics. The "whip" helps make sure that party members are "whipped" into line with their votes (which is to say that they cast their votes in line with party objectives and politics). It takes considerable courage to vote in other ways on highly contested, political matters. As a Congressional Fellow, I was able to help staffers from several offices interpret materials supplied to them. This work made a difference in two areas. I was able to help make sure that no "stupid" embryo bill went forward by providing clear basic information about the science of embryo development. I was also able to inform some decisions about the environment by showing how certain policy options would impact constituents in Arizona in more complex ways than the party leadership understood.

The second lesson I learned was about "acknowledgment" and, more specifically, who gets credit for what. I wrote a number of speeches and opinions, none of which ever had my name on it. This taught me humility. It also allowed me to see that academics are not rewarded enough for working anonymously for the public good or for teamwork. This lesson reminds me to look more closely at the records and contributions of my own faculty members when I am evaluating their work,

[11] J. Maienschein, "Understanding embryos in a changing and complex world: A case of philosophers and historians engaging science," *Erkenntnis* 79: Suppl. 5 (2014), 999–1017. J. Maienschein, *Embryos under the Microscope: The Diverging Meanings of Life* (Cambridge, MA: Harvard University Press, 2014).

and to help the institution recognize a more diverse set of academic contributions that need not include identified authorship.[12]

In addition to working with Congress, I had the good fortune to spend a great deal of time and energy educating federal judges through the Federal Judicial Center's education programs. Under the aegis of this program, I presented seminars and/or keynote lectures at the 9th, 10th, and 11th District courts' annual meetings. For the 9th Circuit Court meeting held in Hawaii, some of the California judges wanted more insight into how others were using genetics in legal cases involving human embryos. Understanding embryos and their role in law and policy made for lively discussions. The 10th Circuit Court met in Santa Fe, New Mexico. The program was designed to include an after-lunch friendly competition. Several of us who would be presenting in the afternoon were invited to give a three- to five-minute preview for our session so as to recruit our audience for the afternoon session. My planned seminar on embryos attracted one of the largest audiences. This experience contrasted markedly with what happened the following year. At the 11th Circuit Court meeting in Florida, the session began with posting of the flag, reciting the Pledge of Allegiance, and other appropriately patriotic rituals. Then, as I was introduced as the opening keynote speaker to discuss the science and social understanding of embryos, a small group walked out in protest. I can only presume that they did not want to hear what I had to say; they already had their intuitions and opinions, and they were engaging in a sort of denial that scientific or historical knowledge could make any difference.

Sometimes, however, luck favors learning. As part of my keynote address at the 11th Circuit Court meeting, I explained the ruling by Judge Royce Lambert of the District of Columbia courts, and I noted that a ruling on an appeal was due any day.[13] As it turns out, the ruling I alluded to was rendered that very day. Later during that meeting as I walked around Disneyworld in Orlando during a scheduled afternoon off, I kept encountering judges who had heard my keynote address. They thanked me for my contribution and said they would not otherwise have paid any attention to news of the Lambert ruling. A few of the judges also acknowledged that while they previously thought they knew that "life begins at conception," they now understood that this was biologically

[12] D. Chubin and J. Maienschein, "Staffing science policy-making," *Science*, 290:5496 (2000), 1501.

[13] For further discussion, see J. Maienschein, *Embryos under the Microscope.*

complex and they could see how this complexity in turn raised challenging and intriguing social, legal, and policy questions. In general terms, then, my goal of educating judges about the subtleties and complexities of science worked, notwithstanding the fact that several judges elected to walk out before I had uttered a word. From this experience, I conclude that scholarly history and philosophy of science studies can help some understand why we need to accept nature as it is and not as some may wish it was.

While I have found it fascinating and gratifying to have had these two types of real-world applied teaching experiences about embryos and society – with congressmen and judges – as I become more senior, I find myself focusing on the fact that one person's work is likely not enough to bring about long-term sustainable change. This brings me to my current work at ASU's School of Life Sciences. I believe that all university faculty have a responsibility to help grow academic institutions to support the kind of use-inspired and engaged work with significant impact that this volume is presenting. One way to do this is through educational programs.

Extending the Impact: The Embryo Project in Action

Graduate students at ASU's School of Life Sciences can study history and philosophy of science through the history and philosophy of science PhD program. Alternatively, they can pursue one (or more) of three tracks within the biology and society master's of science and PhD programs. The first track, history and philosophy of science, explores foundational biological sciences and the social contexts for the historical cases and philosophical analyses. The second track, bioethics, policy, and law, looks at issues at the intersection of these areas of study. The third track is ecology, economics, and ethics of the environment (4E), and it engages policy and ethical study in ways that cut across traditional disciplinary boundaries and go beyond traditional environmental studies.

The students in these graduate programs work together with faculty members to build an innovative educational and research environment with considerable potential for broad social engagement and impact. One example of this is the Embryo Project.[14] The Embryo Project is a

[14] J. Maienschein (director), The Embryo Project, Center for Biology and Society, Arizona State University (2016). Available at: https://cbs.asu.edu/embryo-project. Accessed on: October 28, 2016.

collaborative initiative that engages and connects researchers interested in the history, science, and context of embryos. It has overlapping mandates to pursue and promote university education, research, and public outreach.

The Embryo Project produces the online and open access *Embryo Project Encyclopedia*,[15] which has an ISSN and now receives more than a million page views per year. The use of Facebook, Twitter, Reddit, and other social media outlets has considerably expanded the reach of the *Encyclopedia*, which now reaches a wide international audience. Designed for broad, general audiences (persons who have between a ninth grade and undergraduate education), the *Encyclopedia* is reaching teachers, high school students, and public officials. We know this from the many messages we receive from these audiences. The *Encyclopedia* is also proving to be a useful resource for scholars. We know this from the scholarly citations in articles published in journals like *Science*, *PLOS Biology*, and the *Proceedings of the National Academy of Sciences*, as well as more public outlets.

In 2006, we began the Embryo Project with an initial grant from the National Science Foundation.[16] Without these initial funds, several subsequent grants and the steadfast generous support of ASU President Crow, the Embryo Project would not have been possible. We started with a small cohort of several graduate students and a handful of undergraduate students (many of whom benefited from the very successful *Ask a Biologist*[17] outreach project).

As part of the Embryo Project, we offer a semester-long writers' seminar with a view to helping students learn how to produce content for the *Embryo Project Encyclopedia*. Initially, the seminar was held once a year and about half a dozen students participated. Now we offer the seminar every semester for a dozen or so students. Students apply to take the seminar and most of those admitted take it more than once. The

[15] *Embryo Project Encyclopedia*, Center for Biology and Society, Arizona State University (2015). Available at: http://embryo.asu.edu/. Accessed on: October 28, 2016.

[16] *SGTR: The Embryo Project Training and Research.* Award Number: 0957085. Principal Investigator: J. Maienschein. Co-Principal Investigators: M. Laubichler, C. Norton, M. Dietrich. NSF Organization: Social and Economic Sciences. NSF Program: Science, Tech & Society. Available at: www.nsf.gov/awardsearch/showAward?AWD_ID=0957085&HistoricalAwards=false. Accessed on: October 28, 2016.

[17] *Ask a Biologist*, School of Life Sciences, Arizona State University (2016). Available at: https://askabiologist.asu.edu/. Accessed on: October 28, 2016.

seminar is taught by myself and two graduate students. We each read over 1,000 pages each semester, as we work together through many drafts and revisions.

Each student develops a cluster of articles that connect. The cluster might include articles on a key person, relevant literature, an experiment, or an organization. Topics include anything related to development "from Aristotle up to tomorrow," as I like to put it. The students come up with great ideas and have lively imaginations. We also guide them with lists of "most wanted" articles. All articles must follow the style manuals developed by the graduate students, in particular Erica O'Neil. Some students have written about issues in reproductive biology, including recently a series of articles on the Catholic and Republican Arizona roots for Planned Parenthood and family planning. Others have written on core biological topics, including an excellent cluster on germ layers. One student has written on in vitro fertilization in China, and this article has attracted a lot of attention, as have a number of articles on cellular aging.

At the end of the semester, students submit their portfolio of articles for grading. Stellar submissions are forwarded to Steve Elliott, the editor in chief of the *Embryo Project Encyclopedia*. Elliott has set up a series of protocols for review, fact-checking, and digital publishing and has developed an infrastructure of graduate students to support the editing and content management required for this project. Elliott and Federica Turriziani Colonna also teach an editing and digital publishing course. This course along with the seminar are part of a new nonfiction writing certificate program offered at ASU.

In addition to the writing, reviewing, editing, and digital publishing (all against a background of well-tested protocols and manuals to provide consistency and quality control), Elliott has developed an internship for illustrators. The illustrators are adding images to the *Embryo Project Encyclopedia* to assist with the interpretation of some of the more complex ideas.

As we look toward the future of the Embryo Project, there are two "next steps" on the horizon. First, with National Science Foundation training grants, we have been able to bring international graduate students and scholars into the process, and we hope to "franchise" the Embryo Project so that others can teach their own writing seminars and have their students submit their articles to our editorial team for publication. Second, we are beginning to add features to the Embryo Project that we are developing as part of the Marine Biological

Laboratory History Project.[18] Thousands of photographs from the archives at the Marine Biological Laboratory in Woods Hole, Massachusetts, have been digitized and made available for the *Embryo Project Encyclopedia*, allowing the public to engage with the *Encyclopedia* 's content in new ways.

The content of both of these projects – the Embryo Project and the Marine Biological Laboratory History Project – reside on servers in the ASU library, and the entries are all part of the HPS (History and Philosophy of Science) Repository.[19] All of this is coordinated and overseen by graduate student and project coordinator Kate MacCord.

These two projects are oriented toward giving every student a set of skills and experiences with communicating science to multiple audiences and with digital publishing. Students also take turns giving presentations to the public through regular events at ASU, the Arizona Science Center, ASU Homecoming, and other venues in Arizona and nationally. Our group has helped develop outreach programs for the History of Science Society[20] and Philosophy of Science Association,[21] in particular, serving as the organizers for a Joint Caucus for Socially Engaged Philosophers and Historians of Science.[22] Our students also help run the international Digital HPS group,[23] which is oriented toward sharing results including about how to present ideas to a broader public. In all cases, the students serve as the organizers and leaders, and they have developed a series of manuals and protocols for all aspects of the work. These are designed to share the skills and tools, and not just the final products, with anybody – in a completely openly accessible and open-sourced way. These students will graduate and others will take their places in this lively training program that engages students in effective public outreach and advocacy for understanding science in its historical and philosophical context.

18 Marine Biological Laboratory History Project, the University of Chicago (2016). Available at: http://history.archives.mbl.edu. Accessed on: October 28, 2016.

19 History and Philosophy of Science Repository, Arizona State University (2015). Available at: http://hpsrepository.asu.edu. Accessed on: October 28, 2016.

20 History of Science Society, University of Notre Dame (2016). Available at: http://hssonline.org/. Accessed on: October 28, 2016.

21 The Philosophy of Science Association (2016). Available at: http://philsci.org/. Accessed on: October 28, 2016.

22 Joint Caucus of Socially Engaged Philosophers and Historians of Science (2016). Available at: http://jointcaucus.philsci.org/. Accessed on: October 28, 2016.

23 Digital HPS. Available at: http://digitalhps.org. Accessed on: October 28, 2016.

Conclusion

I remain amazed at my good fortune in being able to work at an institution that values ethics-with-impact work that is at the intersection of scholarship, advocacy, and public good. ASU's charter captures this beautifully:

> ASU is a comprehensive public research university, measured not by whom we exclude, but rather by whom we include and how they succeed; advancing research and discovery of public value; and assuming fundamental responsibility for the economic, social, cultural and overall health of the communities it serves.[24]

Many support this new university model that embraces education, learning, and applying knowledge for the benefit of faculty, students, and the community as a whole. Among the many supporting voices is that of former Columbia University provost Jonathan Cole. At a recent event at ASU, Cole held that Ivy League schools have become predictable and their admission process is so exclusive that "the students are 'boring.' They all have perfect scores. They've never deviated from the beaten path."[25] Although Cole's entire education and career has been at Columbia, he said, "The Ivy League is not where it's at."[26] But is the ASU model transposable to other places? A study by Jon Marcus for the Hechinger Report, which appeared in *U.S. News and World Report*, asks this provocative question.[27] As Cole puts it, we have to make changes because "[m]ost education and most research is taking place at state universities, and we cannot let them fail."[28]

Inspired by the other stories brought to life in this book, we can all work together to extend the institutional as well as personal support for carrying out scholarship with impact. Individual actions are important, and even necessary. But actions that extend beyond the power of one, especially through education and teamwork, are especially important.

[24] ASU, *Arizona State University Charter.*

[25] Quoted in: M. B. Faller, "ASU setting path 'Toward a More Perfect University,'" *ASU Now: Access, Excellence, Impact* (March 11, 2016). Available at: https://asunow.asu.edu/20160311-solutions-asu-setting-path-toward-more-perfect-university. Accessed on: October 28, 2016.

[26] Quoted in: M. B. Faller, "ASU setting path."

[27] J. Marcus, "Arizona state president: School is model of the new U," *U.S. News and World Report* (March 17, 2015). Available at: http://www.usnews.com/news/college-of-tomorrow/articles/2015/03/17/arizona-state-president-school-is-model-of-the-new-u. Accessed on: October 28, 2016.

[28] Quoted in: M. B. Faller, "ASU setting path."

The Embryo Project and ASU provide examples of a firm commitment to supporting the next generation of change makers by helping them learn to communicate effectively with multiple audiences.

Acknowledgment

Thanks to the National Science Foundation for a series of grants and to Arizona State University's president's office for continued support. Special thanks to the Embryo Project team led by Kate MacCord, Steve Elliott, Erica O'Neil, Federica Turriziani Colonna, Valerie Racine, Alexis Abboud, Kelle Dhein, Karen Wellner, and Mary Sunderland. And special thanks to Jessica Ranney.

9

The Flint Water Crisis

ARON SOUSA

By and large, this anthology focuses on stories of harm accruing to medical patients and medical research subjects. This chapter is different in that it is about public health – or, more specifically, about a public health crisis that might never have happened and that might never have become the tenacious national story it did: the Flint Water Crisis. The story of the Flint Water Crisis demonstrates, at least, this: that basic health determinants like wholesome food, safe water, and public safety depend on rational, scientific decisions made in the public's best interest; that when bad decisions are made, any of us can suddenly find ourselves becoming patients and public health research subjects; that the fate of our health is tied to decisions of communal importance made by everyone involved in public health, from state governors to water supply staff; and that media attention helps, particularly in terms of rooting out the facts about government dealings, but it doesn't fix the problems. Perhaps most importantly, the story of the Flint Water Crisis shows that fundamental democratic principles like enfranchisement, academic freedom, a vibrant free press, and government workers' fiduciary responsibilities can have direct consequences for people's health.

The crisis is explained in more detail below, but here is the basic story. Decades of economic distress in the city of Flint, Michigan – a predominantly African American city with longstanding, Democratic Party leadership – led finally to the elected Republican leadership in the state of Michigan appointing an emergency manager to take over running the city. To save costs, Flint's water supply was switched over to a different source and was then mismanaged in a way that caused significant amounts of lead to leach into drinking water in many parts of the city.[1]

[1] For a relatively complete timeline of the Flint Water Crisis, see S. Roy, "Michigan Governor Rick Snyder's office released this Flint water timeline to address questions on who knew what and when," *Flint Water Study Updates* (January 19, 2016). Available at: http://flintwaterstudy.org/2016/01/michigan-governor-rick-snyders-office-released-this-

By the time the state and national governments admitted that local activists and academic researchers were right – that Flint was harming its people with unwholesome water – some people had already accrued lasting harm.[2]

The Flint water debacle comes as the latest insult and existential crisis for an American "rust-belt" (postindustrial) city with a shrinking population consisting of a majority of minorities. As I write, in early 2016, the water in Flint has put thousands of people at risk of exposure to lead, a known neurotoxin, and likely also caused an outbreak of legionella pneumonia with resultant deaths.[3] Beyond the health risks of lead and legionella, the people of Flint have had their lives turned upside down by the water crisis. They are forced to drink bottled water – something that sounds trivial until you have to try, particularly as a poor person in a city with large amounts of blight and infrastructure collapse, to get enough water every day to drink, cook, and bathe. For nearly two years now, the people of Flint have faced the struggle and hassle of hauling their drinking and cooking water from fire stations and food pantries, and there is no end to this situation in sight.[4] By 2016, the people of Flint had little trust in a system that has left them disenfranchised and had little trust in a supposed public safety net that failed to find and announce health hazards.[5] Indeed, in 2015–2016 I was in meetings with governmental leaders ranging from local and state officials to the US Surgeon General and the US Secretary of Health and Human Services where the main discussion was how to convince the people of Flint to trust government assurances of water safety.

flint-water-timeline-to-address-questions-on-who-knew-what-and-when/. Accessed on: May 5, 2016. See also A.M. Barry-Jester, "What went wrong in Flint," *FiveThirtyEight* (January 26, 2016). Available at: http://fivethirtyeight.com/features/what-went-wrong-in-flint-water-crisis-michigan/. Accessed on: May 5, 2016.

2 A. Goodnough, "Legionnaires' outbreak in Flint was met with silence," *New York Times* (February 22, 2016). Available at: www.nytimes.com/2016/02/23/us/legionnaires-outbreak-in-flint-was-met-with-silence.html. Accessed on: October 18, 2016.

3 See A. Goodnough, "Legionnaires' outbreak."

4 M. Dolan, "Far more Flint homes have lead lines than expected, report shows," *Detroit Free Press* (September 29, 2016). Available at: www.freep.com/story/news/local/michigan/flint-water-crisis/2016/09/28/more-than-half-flint-homes-could-have-lead-lines-report-shows/91225284/. Accessed on: October 21, 2016.

5 P. Egan, "Poll: Flint residents don't trust water filters," *Detroit Free Press* (June 2, 2016). Available at: www.freep.com/story/news/local/michigan/flint-water-crisis/2016/06/02/poll-flint-residents-dont-trust-water-filters/85247496/. Accessed on: October 19, 2016.

At the time of the Flint Water Crisis I was interim dean of Michigan State University's College of Human Medicine. I don't live in Flint – I live in East Lansing, the home of the main campus of Michigan State University, a predominantly white city about an hour west of Flint – and by any standard I have been a minor player in the Flint Water Crisis. But I have worked with people in the community for years, and I have been involved in the response to the crisis as an administrator trying to support, guide, and protect faculty of the College of Human Medicine who have worked to enact change. As detailed below, I led in the establishment of Michigan State University's public health program in Flint several years before the water crisis, and so before and throughout the Flint Water Crisis, my college has had medical and public health students in Flint as well as a core group of public health faculty who helped bring the facts of this public health emergency to light.

Today my team is helping to bring resources that might help the people of Flint, while also helping researchers and policymakers learn more about lead poisoning, early childhood education, and how to improve access to safe food and water. But until recently, bringing the emergency to light was pretty much all any faculty of the College of Human Medicine could do, because government officials, some of whom actively misrepresented the facts of the water contamination, kept claiming there was no problem.[6] This created a very challenging situation for us. University-based public health programs like ours are not public health departments nor environmental protection agencies; university-based or -affiliated educators, physicians, researchers – even our university president – can't order city-wide water treatments, replacement of pipes, mass delivery of clean water, or declare a public health emergency. Only the government can do that. While the government was in denial, nothing very effective could be done. As a consequence, for many months, all medical and public health faculty could do was try to find and show the truth, and try to take care of the people being harmed.

My goal in this chapter is to provide a greater understanding of what has happened in this crisis and to highlight the importance of individual action in public health, whether that be by government workers, elected officials, university administrators, or university-affiliated researchers and healthcare professionals. To me, the most interesting question raised

[6] S. Roy, "MDEQ mistakes and deception created the Flint Water Crisis," *Flint Water Study Updates* (September 30, 2015). Available at: http://flintwaterstudy.org/2015/09/commentary-mdeq-mistakes-deception-flint-water-crisis/. Accessed on: October 18, 2016.

by Flint disaster is: Why did governmental systems designed to protect people fail and fail again in this case? I would summarize the answer to this provocative question in this way: failures to listen, failures to engage, failures to stand up. It appears that administrators at the state and federal levels assumed that their systems would work and did not welcome other opinions, especially those that said the system was endangering the lives of children.[7] It also appears that scientists and employees within the system who brought bad news were denied, shut down, and sometimes perhaps even punished.[8] More generally, the government agencies charged with dealing with water purity had no system for hearing bad news in a way that would move them to act appropriately.

The only thing that turned this around was unrelenting media attention to the work and claims of activists and whistleblowers like LeeAnn Walters, Mona Hanna-Attisha, Miguel Del Toral, Marc Edwards, and Melissa Mays.[9] Unlike many public health travesties, this one became a persistent focus of the American media, no doubt in part because of the basic expectation that your babies should be able to safely drink water from the tap. Based on my conversations with activists in Flint and people in public health in government, the Black Lives Matter movement – a movement that successfully focused national attention on brutal treatment of African Americans at the hands of police officers across the United States – raised the consciousness of many white Americans about racial disparities experienced at the hands of government agencies. As the Flint Water Crisis finally became a national story, we heard about many more cities in America facing similar disasters.[10]

7 Associated Press, "Doctors urge Flint to stop using water from Flint River," *Crain's Detroit Business* (September 28, 2015). Available at: www.crainsdetroit.com/article/20150928/NEWS01/150929872/doctors-urge-flint-to-stop-using-water-from-flint-river%20Accessed%20October%2018,%202016. Accessed on: October 18, 2016.

8 Sarah Hulett, "High lead levels in Michigan kids after city switches water source," *All Things Considered* (September 29, 2015). Available at: www.npr.org/2015/09/29/444497051/high-lead-levels-in-michigan-kids-after-city-switches-water-source.html. Accessed on: October 19, 2016.

9 A. Goodnough, M. Davey, and M. Smith, "When the water turned brown," *New York Times* (January 23, 2016). Available at: www.nytimes.com/2016/01/24/us/when-the-water-turned-brown.html. Accessed on: October 19, 2016.

10 O. Milman and J. Glenza, "At least 33 U.S. cities used water testing 'cheats' over lead concerns," *The Guardian* (June 2, 2016). Available at: www.theguardian.com/environment/2016/jun/02/lead-water-testing-cheats-chicago-boston-philadelphia. Accessed on: October 19, 2016.

Organizing Attention to Flint's Public Health Needs

Michigan State University's College of Human Medicine has campuses all over the state of Michigan, and we try to tailor our local activities to local needs. Our college is a child of the 1960s, founded in 1964 with a motto of "serving the people," a phrase borrowed from Chairman Mao. Our founding dean, Andrew Hunt, envisioned a new kind of medical school tightly linked to the communities in which students would learn.[11] Ours is a community-based medical school in that the students' third- and fourth-year clinical training is done at seven campuses scattered throughout our state. The model was designed to deepen students' connections to communities, to better educate students, and to retain doctors in underserved rural and urban communities.[12] This model has been successful enough that nearly all American medical schools now have some community-based education opportunity for students.

The community-based model also guides many of our researchers' interests. The model followed closely on the founding heritage of Michigan State University as the first US land-grant college.[13] In 1855, the people of the state of Michigan set aside a large tract of land to create a university dedicated to improving the lives of its citizens; first called Michigan Agriculture College, this university is now Michigan State University. Like the other land-grant universities that would follow, Michigan State University first focused on agriculture and technical education. Michigan State University's modern extension service still has employees in each county of the state, but now 60 percent of their work is related to health, mostly around nutrition. Our college's placing of our medical students and faculty in communities around our state has always been understood as an extension of the land-grant tradition of our university. This is important because it means our institution has philosophically – and institutionally – empowered all our faculty to serve the people immediately around them. While all

[11] A.D. Hunt, *Medical Education, Accreditation, and the Nation's Health: Reflections of an Atypical Dean* (East Lansing, MI: Michigan State University Press, 1991).

[12] A.L. Wendling, J. Phillips, W. Short, C. Fahey, and B. Mavis. "Thirty years training rural physicians: Outcomes from the Michigan State University College of Human Medicine Rural Physician Program," *Academic Medicine*, 91:1 (2016), 113–119.

[13] "The Nation's pioneer land-grant university," Michigan State University (2016). Available at: https://msu.edu/morrill-celebration/history.html. Accessed on: October 19, 2016.

universities value national-level research and service, we also highly value locally relevant research and service. We often hire people we believe will do locally focused work.

About a decade before the Flint Water Crisis, I started working to build a public health program in Flint and to expand our medical education presence there. I did so in my capacity as senior associate dean for academic affairs of the College of Human Medicine (the position I had from 2005 to 2015, before being named interim dean). Beginning in early 2006, Flint community leaders came to our college with an interest in an expanded medical school to fit their changing economic strategies, shifting focus from the traditional manufacturing base of Flint to focusing on healthcare and education. One of the opportunities presented was to expand on the forty-year history of our medical college in Flint. Over the course of about six years, community leaders and college leadership (including myself) discussed opportunities ranging from creating a new medical school to simply adding a few more clinical students in Flint. We ultimately arrived at a plan to move the public health program of the college from East Lansing to Flint. In our planning and grant-writing, college administrators explained our desire to create a public health unit dedicated to community-participatory public health research. With the support of about $12 million in grants from the philanthropic C. S. Mott Foundation and another $4 million in support from the main city hospitals (Genesys Health, McLaren-Flint, and Hurley Medical Center), the college set about hiring a group of community participatory public health researchers in 2014.

The college's first step was to set up a community advisory board including representatives from stakeholder institutions as well as genuine, grassroots, community-based organizations. This advisory board came up with three areas of research focus: chronic diseases, behavioral health, and healthy behaviors. As well, the advisory board said it wanted researchers who focused on interventions that would decrease racial and ethnic disparities in healthcare in Flint. The African American community-based organizations in particular were tired of academics coming to Flint, taking their data, "finding" problems that the community already knew they had, and then leaving for their home institutions. The community representatives wanted a better, more inclusive research experience. Our college's recruitment of faculty to Flint was remarkably successful. We attracted high-level researchers from Brown University, Johns Hopkins University, and the University of Michigan's public health school.

The Economic and Political Roots of the Crisis

The vulnerability of the Flint community to significant public health problems is a result of economic hardship and migration patterns over the last forty years. During the heyday of the American auto industry and strong unions, Flint was an exemplar of rising dreams of the middle class, including for African American families. Many were able to buy homes, see their children do well, and look forward to safe retirements. But with the collapse of the American auto industry, Flint has now shrunk from a city of 200,000 people with one of the highest standards of living in the country in the 1960s to a city of under 100,000 people with some of the lowest quality-of-life and health standards in the country.[14] Today, Flint suffers from high rates of poverty and crime. The city has been beleaguered for decades and, as a practical matter, has struggled even to support a major grocery store.[15]

In many neighborhoods – particularly those that have now been found to have high water lead levels and increased numbers of children with high blood lead levels – once-fine homes sit abandoned and others are surrounded by empty yards where vacated houses have been demolished. On top of everything else, Flint was especially hard hit by the foreclosure crisis of the early 2000s, caused after many questionable mortgages were pushed on unsuspecting homeowners.[16] The infrastructure of these neighborhoods is poor and old, and lead pipes linking houses to city mainlines are the rule rather than the exception. With few occupied homes on a given block, there can be little water usage, and so water sits in mainlines and service lines with plenty of time to corrode pipes, consume and deplete disinfectants like chlorine, and allow bacterial growth.

The combination of Flint's shrinking tax base, long-term debt load, crumbling infrastructure, and population in high need of social services has crushed the city under extreme financial weight for many years. Starting in 2011, Michigan's Governor Rick Snyder appointed a series

[14] T. Lee, "The rust belt: Once mighty cities in decline," *MSNBC* (undated). Available at: www.msnbc.com/interactives/geography-of-poverty/ne.html. Accessed on: October 20, 2016.

[15] A. Allington, "Grocery exodus has Flint shopping for answers," *Marketplace* (May 22, 2015). Available at: www.marketplace.org/2015/05/22/business/grocery-exodus-has-flint-shopping-answers. Accessed on: October 18, 2016.

[16] C. Wilson, "The problems of legacy cities," *Michigan State University Extension* (September 21, 2016). Available at: http://msue.anr.msu.edu/news/the_problems_of_legacy_cities. Accessed on: October 20, 2016.

of emergency managers to tackle the problem of managing Flint's needs in the severe economic crisis situation. These managers have had the power to executively overrule Flint's elected mayor and city council.[17] Throughout Michigan, there are state-installed emergency managers in various cities, including Detroit. And as many reporters and political commentators have noted, almost all of those cities are majority African American.[18] As such, it is not just the appearance that African Americans have been more likely to be democratically disenfranchised in Michigan, but a statistical fact.

By effectively losing the power of the local ballot box through the appointment of an emergency manager, the people of Flint lost their ability to give or withdraw their consent to the public health decisions determining the wholesomeness of their water. The lack of enfranchisement and access to a meaningful ballot box fundamentally changes the relationship of people to their government and vice versa.

The emergency manager system primed the people of Flint to be suspicious of their government when it came to their water supply (and just about everything else), but the African Americans among them were also primed by their knowledge of the US Public Health Service's forty-year study of untreated syphilis in Tuskegee. The basic trust that most people have in their government to provide for the general welfare is undercut among many African Americans by the felt experience of Tuskegee.[19] That experience has been replayed in the lives of the residents of Flint through the water crisis. I have attended town halls of pitched fury with accusations of attempted genocide and purposeful poisoning by the government. The people of Flint are very angry and are more than willing to entertain the notion that government has been out to get them.

So how exactly did the crisis happen? In 2010, several counties surrounding Flint decided to create a new water district, called the Karegnondi Water Authority. The Karegnondi Water Authority would use water piped directly from Lake Huron, the nearest Great Lake, to provide a cheaper source of drinking water than what was provided by the Detroit Water and Sewerage Department, from which Flint had been

[17] J. Bosman and M. Davey, "Anger in Michigan over appointing emergency managers," *New York Times* (January 22, 2016). Available at: www.nytimes.com/2016/01/23/us/anger-in-michigan-over-appointing-emergency-managers.html. Accessed on: October 18, 2016.

[18] See J. Bosman and M. Davey, "Anger in Michigan."

[19] S.M. Reverby, *Examining Tuskegee: The Infamous Syphilis Study and Its Legacy* (Chapel Hill, NC: University of North Carolina Press, 2009).

drawing its water. That effort was going to take several years to complete, and in the meantime, in April 2014, Flint's elected leaders and governor-appointed emergency manager decided to switch Flint's drinking water from the Detroit Water and Sewerage Department to the Flint River in order to save about $5 million over the years it was expected to take to complete the Karegnondi Water Authority connection between Flint and Lake Huron.[20]

It was the switch to the Flint River as a source of water that caused the Flint Water Crisis. The Flint River, like many rivers, has relatively corrosive water with many minerals and chlorides from street runoff, from percolation of substances through the soil, and from minerals and metals in the riverbed. By contrast, Lake Huron is huge, and the water has much less contact with minerals and chlorides, and so is much less corrosive.

If the change in water source to the Flint River had been well managed technically, much less harm would have come to the people of Flint. But – perhaps again to save money? – for some reason the water was not properly treated. The highly corrosive water drew lead out of old service lines and into the drinking water and then into the bloodstreams of thousands of people in Flint. People in Flint drank the water, cooked with it, and used it to mix with baby formula. The water from the Flint River often was discolored, smelled bad, and tasted bad, which made people immediately suspicious about its safety. People who could afford filters and bottled water switched on their own, while those who could not afford filters and bottled water made do.

There were many complaints by citizens about the quality of the water, and at its Flint plant, the automaker giant General Motors switched itself back to the Detroit Water and Sewerage Department because the Flint River water was so corrosive it was rusting car parts.[21] I know from my discussions with some hospital executives that they also knew there was something wrong with the water in the first week of the switch to the Flint River, because the steam-based autoclaves were corroding the surgical instruments they were meant to sterilize.

[20] See S. Roy, "Michigan Governor Rick Snyder's office released," and A.M. Barry-Jester, "What went wrong in Flint."

[21] M. Colias, "How GM saved itself from Flint water crisis: Rusting engine blocks flagged big problem," *Automotive News* (January 31, 2016). Available at: www.autonews.com/article/20160131/OEM01/302019964/how-gm-saved-itself-from-flint-water-crisis. Accessed on: October 20, 2016.

By March 2015, Flint's city council voted to move the city's water source back to the Detroit Water and Sewerage Department.[22] But the state-appointed emergency manager declined, calling the city council's decision "incomprehensible."[23] The local elected officials had no ability to overturn the state-appointee's decision.

Activists Collect Data and Demand Action

Even as state-level officials refused to address local concerns, a group of Flint citizen activists attempted to address the quality of the water and health concerns stemming from the water. Most notable among the activists working at this time was Lee Ann Walters. Walters had her home's tap water tested by the Flint Water Department, and it was found to be very high for lead. Her children also tested high for lead, higher than they had tested before the switch to the Flint River water. At this point, Walters stopped using Flint tap water in her home. She also started trying to figure out what was happening in the rest of Flint. She contacted experts and sorted out that the Flint water was not being treated with anticorrosives.[24] Meanwhile, many local individuals and groups such as Concerned Pastors for Social Action and Community Based Organization Partners (among others) held protests, spoke to City Council, and contacted various politicians.

Walters also decided to contact Miguel Del Toral, a US Environmental Protection Agency (EPA) expert on lead and copper testing. He personally investigated the situation and produced a critical memo, dated June 24, 2015, with a detailed timeline showing city and state actions and inactions. In his memo, Del Toral concluded that the treatment of Flint's water did not comply with the law and that the state's water testing was proving inadequate for public safety.[25]

22 See S. Roy, "Michigan Governor Rick Snyder's office released," and A.M. Barry-Jester, "What went wrong in Flint."

23 R. Fonger, "Emergency manager calls City Council's Flint River vote 'incomprehensible,'" *MLive News* (March 24, 2015). Available at: www.mlive.com/news/flint/index.ssf/2015/03/flint_emergency_manager_calls.html. Accessed on: May 5, 2016.

24 See S. Roy, "Michigan Governor Rick Snyder's office released," and A.M. Barry-Jester, "What went wrong in Flint."

25 M.A. Del Toral, "High lead levels in Flint, Michigan – Interim report," Memorandum from the United States Environmental Protection Agency (June 24, 2015). Available at: http://flintwaterstudy.org/wp-content/uploads/2015/11/Miguels-Memo.pdf. Accessed on: May 5, 2016.

Del Toral's memo was conclusive and convincing, but neither the EPA nor the Michigan's Department of Environmental Quality made effective use of the memo. Susan Hedman, the head of EPA District 5, which includes Flint, told Michigan officials that Del Toral's report was not finalized and was preliminary.[26]

When Del Toral sent his memo to Michigan's Department of Environmental Quality, he also sent it to Lee Ann Walters. His memo included information about her house and her children, so that might have been his reasoning for sending her a copy. But, of course, copying Waters also essentially ensured the memo would become public, an interesting preemptive move by a civil servant. Hedman tried nevertheless effectively to block Del Toral by telling the state the memo was not finalized or official and by apologizing for how the memo was released.[27] (Hedman later resigned in the wake of publicity over these events.)

The Michigan Department of Environmental Quality went on the offensive, attempting to refute the memo, and the spokesperson for the department, Brad Wurfel, called Del Toral "a rogue employee." (Wurfel also resigned.[28])

While Walters was working with Del Toral, she also contacted Marc Edwards, of Virginia Polytechnic Institute and State University, another land-grant university, known as Virginia Tech. Edwards is a water engineer who had taken on the EPA over high lead levels in the water of Washington, DC, in the early 2000s.[29] Edwards subsequently set up the Flint Water Study in his Virginia Tech laboratory and began a citizen science campaign to sample and test Flint water.[30] The people of Flint became deeply engaged with Edwards in the sampling of Flint's water.

[26] Memo from Susan Hedman to Dayne Walling dated July 1, 2015 (and follow-on correspondence). Available at: https://oversight.house.gov/wp-content/uploads/2016/02/7-1-7-2-Emails-from-Hedman-to-Mayor-Walling-of-Flint.pdf. Accessed on: April 5, 2017.

[27] L. Smith, "After blowing the whistle on Flint's water, EPA 'rogue employee' has been silent. Until now," *Michigan Radio* (January 21, 2016). Available at: http://michiganradio.org/post/after-blowing-whistle-flints-water-epa-rogue-employee-has-been-silent-until-now#stream/0. Accessed on: May 5, 2016.

[28] L. Smith, "After blowing the whistle on Flint's water."

[29] A. Halsey III and M. DeBonis, "Water in thousands of D.C. homes might still be contaminated by lead, CDC says," *Washington Post* (December 2, 2010). Available at: www.washingtonpost.com/wp-dyn/content/article/2010/12/01/AR2010120107286.html. Accessed on: June 10, 2016.

[30] M. Edwards, S. McElmurray, A. Pruden, and D. Reckhow, "Updates on special Flint sampling: Flushing and lead, disinfection by-products and Legionella sampling (May 31st 2016)," *Flint Water Study Updates* (May 31, 2016). Available at: http://flintwaterstudy.org/. Accessed on: June 10, 2016.

Edwards categorizes the people of Flint as his collaborators and says they are among the best scientific collaborators he has had. His Flint partners made sure there was sampling for every neighborhood, designed systems to ensure the water samples were not contaminated or tampered with, and returned a much higher number and higher percentage of samples than for any other study the state did.[31]

By September 2015, Edwards and his Flint citizen colleagues were able to demonstrate conclusively that Flint's water was contaminated with lead.[32] Nevertheless, the Michigan Department of Environmental Quality continued to claim that the water was meeting EPA standards.[33] Department of Environmental Quality officials even impugned the integrity of Edwards. The Michigan Department of Environmental Quality's spokesman, Wurfel, wrote the following to the local newspaper, the *Flint Journal*:

> When I said we were unsure how the Virginia Tech team got its results, that's not the same as being surprised that they got them . . . this group specializes in looking for high lead problems. They pull that rabbit out of that hat everywhere they go. Nobody should be surprised when the rabbit comes out of the hat, even if they can't figure out how it is done.[34]

In the next few weeks pediatrician-researcher Mona Hanna-Attisha and her colleagues were able to show that children in Flint were developing toxic levels of lead in their blood.[35] Hanna-Attisha practices at Flint's Hurley Medical Center and is also an assistant professor at Michigan State University's College of Human Medicine, my college. (We had engaged her as faculty several years earlier.) Like Edwards, she soon moved to openly fight governmental official claims that the water was safe.

[31] M. Edwards, "How Jonathan Baldwin Turner saved Flint, Mich.: Public-inspired science and the modern land-grant university," *WKAR* (February 26, 2016). Available at: http://video.wkar.org/video/2365669527/. Accessed on: May 5, 2016.

[32] S. Roy, "Our sampling of 252 homes demonstrates a high lead in water risk: Flint should be failing to meet the EPA Lead and Copper Rule," *Flint Water Study Updates* (September 8, 2015). Available at: http://flintwaterstudy.org/2015/09/our-sampling-of-252-homes-demonstrates-a-high-lead-in-water-risk-flint-should-be-failing-to-meet-the-epa-lead-and-copper-rule/. Accessed on: May 5, 2016.

[33] See S. Roy, "MDEQ mistakes and deception."

[34] S. Roy, "MDEQ mistakes and deception."

[35] R. Fonger, "Elevated lead found in more Flint kids after water switch, study finds," *MLive* (September 24, 2015). Available at: www.mlive.com/news/flint/index.ssf/2015/09/study_shows_twice_as_many_flin.html. Accessed on: October 20, 2016.

The American Academy of Pediatrics recommends that children be assessed for possible lead exposure regularly until age six.[36] As a result, many children in the United States are screened for lead by having blood drawn and tested. Hanna-Attisha studied data from the Hurley Medical Center's own laboratory and found that the number of children with high lead levels in their blood was increasing. She was careful to say that her study did not at that time link the lead levels in water to the increases in children's blood lead levels, but she noted that the water was what had most obviously changed in the environment in the interim.

The Michigan Department of Environmental Quality and the Michigan Department of Health and Human Services claimed that they could not duplicate Hanna-Attisha's results.[37] Just as when Edwards brought forth his results and the Department of Environmental Quality's Wurfel had impugned Edwards's integrity, when Hanna-Attisha brought forth her results, Wurfel called the Flint water situation "near hysteria."[38]

As it turns out, the state-level environmental experts were probably not coding the blood lead level data correctly and were including children from outside Flint in the "Flint data." Within a couple of weeks of the data dispute going public, the state confirmed Hanna-Attisha's findings. The city and state managed to change the system back to Detroit water.[39] Wurfel even apologized to Hanna-Attisha.[40] The city and state began counseling people to use filters or bottled water, although a long-term lead solution remains elusive. One problem is that the corrosive water caused lead to leach into the whole system, and so leaching lead is still present in the pipes and thus still in the water. Switching back to the Detroit River has not simply undone the problems.

And lead is not the only problem resulting from the water mismanagement. The highly corrosive water and the metals it solubilized also

36 "Recommendations for preventive pediatric health care," American Academy of Pediatrics and Bright Futures (May 2015). Available at: www.aap.org/en-us/Documents/periodicity_schedule_oral_health.pdf. Accessed on: May 5, 2016.

37 R. Fonger, "State says data shows no link to Flint River, elevated lead in blood," *MLive* (September 24, 2015). Available at: www.mlive.com/news/flint/index.ssf/2015/09/state_says_its_data_shows_no_c.html. Accessed on: October 20, 2016.

38 Associated Press, "Doctors urge Flint to stop using water from Flint River."

39 "Events that led to Flint's water crisis," *New York Times* (undated). Available at: www.nytimes.com/interactive/2016/01/21/us/flint-lead-water-timeline.html. Accessed on: October 20, 2016.

40 R. Erb, "Flint doctor makes state see light about lead in water," *Detroit Free Press* (October 12, 2015). Available at: www.freep.com/story/news/local/michigan/2015/10/10/hanna-attisha-profile/73600120/. Accessed on: October 20, 2016.

consumed the chlorine used to disinfect the water. According to Edwards of Virginia Tech and his Flint Water Project, about 80 percent of area homes had inadequate chlorine for disinfection of drinking water at their taps during the height of the crisis. Temporally concomitant with the switch in water sources, Flint had an outbreak of legionella pneumonia, a type of bacteria that can cause serious pneumonia if it is inhaled in microscopic water droplets, like those that come from showers and air conditioning systems. The legionella pneumonia outbreak is one of the largest per capita in US history, and ten people are known to have died from it by early 2016.[41] During the height of the Flint Water Crisis, before the legionella outbreak became very visible, there were two orders issued to boil drinking water to kill bacteria. While boiling will kill legionella and other water-borne bacteria, unfortunately, boiling water only serves to increase the concentration of lead in drinking water.

Why the Failure of Protections?

The water plan all along had been to move from the relatively pricey Detroit water system to a new, lower cost pipeline that would draw water directly from Lake Huron. Over the years Flint had occasionally used Flint River water as a stop-gap water source if there was a short-term interruption in the flow of Detroit water, but this switch was specifically planned to be for several years, which was much longer than Flint had ever used the Flint River. As months passed following the switch to the Flint River, the Michigan agencies charged with tracking lead levels and blood lead levels watched as these levels increased with time but did not take action because the levels did not reach "action levels" set by the EPA.[42] These scientists appeared not to realize that the increases were statistically significant – indeed, alarming – in part because they did not realize the geographic relationship of the water lead levels with children with elevated blood lead levels.

Clearly, there was a scientific and engineering failure in the design and implementation of the switch from the Detroit water to the Flint River source in April 2014. Like most systematic failures, the series of events

[41] A. Pruden and O. Schawake, "Legionnaire's disease possibly associated with Flint River water supply," *Flint Water Study Updates* (January 15, 2016). Available at: http://flintwaterstudy.org/2016/01/legionnaires-disease-possibly-associated-with-flint-river-water-supply/. Accessed on: May 5, 2016.

[42] See S. Roy, "MDEQ mistakes and deception."

that lead to children with high blood lead levels in Flint represents a litany of bad decisions, one after another. At this point it is not clear why the city's water department did not provide anticorrosive treatment when it switched to Flint River water. And it is not clear why the Michigan Department of Environmental Quality did not make the water plant follow the standard practice of anticorrosive treatment required by law. Further, it is not clear why the US EPA also did not require anticorrosive treatment even though it had an employee, Miguel Del Toral, doing his job, writing in April 2015 to his superiors about how dangerous the lack of corrosion controls in Flint were.

It is clear from public statements and publicly released emails[43] that the leadership of the state, the Michigan Department of Environmental Quality, the regional EPA, and the governor-appointed Flint emergency manager were all dismissive of the concerns of Flint residents. When physicians and scientists supported the claims of residents with data, they were also dismissed as "rogue" (as Miguel Del Toral was called), out to find lead anywhere (as Marc Edwards was accused), "wrong" and "near-hysteri[c]" (terms applied to the work of Mona Hanna-Attisha). When the EPA scientist Del Toral told the regional manager of the EPA that the Flint River water was not being adequately treated, the EPA manager deferred to the state of Michigan, and state officials then referred to Del Toral's memo as a memo written by a rogue employee.[44] When engineer-researcher Edwards of Virginia Tech publicly said the water was not being adequately treated and demonstrated through community water sampling that the blood levels were high, state officials derided *him* as a maverick.[45] When the pediatrician-researcher Hanna-Attisha reported elevated numbers of children testing high for blood lead levels, state officials denied *her* findings and said that she was wrong.[46]

It is not hard to understand why others in the scientific community did not stand up when they should have. A federal government scientist (Del Toral) was effectively sidelined in deference to the state, and

43 S. Roy, "MDEQ mistakes and deception."

44 Lindsey Smith, "The mom and the EPA 'rogue employee' who exposed Flint's water crisis," *Public Radio International* (January 23, 2016). Available at: www.pri.org/stories/2016-01-23/after-blowing-whistle-flints-water-epa-rogue-employee-has-been-silent-until-now. Accessed on: October 20, 2016.

45 D. Hohn, "Flint's water crisis and the 'troublemaker' scientist," *New York Times* (August 16, 2016). Available at: www.nytimes.com/2016/08/21/magazine/flints-water-crisis-and-the-troublemaker-scientist.html. Accessed on: October 18, 2016.

46 See R. Erb, "Flint doctor makes state see light."

academics were being publicly derided for publicizing their findings. It is now clear that some University of Michigan scientists decided not to investigate the water issue in Flint because they wanted to work with the state rather than against it.[47] (The University of Michigan is a separate institution from Michigan State University.) From my own conversations, I know that some administrators in the hospitals and at my own university were worried about angering Michigan's governor and state legislature by publicly taking the concerns of Flint residents seriously.

Even at Michigan State University, where Hanna-Attisha works with me, we were a little slow to recognize the failures of the state. As late as September 2015, we simply did not believe that the state scientists and administrators would be so wrong about the levels of lead in children in Flint. We surely knew the water tasted bad and looked bad, but those of us who did not live or work in Flint struggled to believe that the scientists at the state level would let the water be dangerous. It simply never occurred to us that state-level epidemiologists would not be able to accurately use zip codes to log the data. And, to be honest, I think many of us thought that the concerns of Flint citizens might be overblown. As in many cities, there have been contingents of Flint activists who have consistently opposed many attempts at development or change in the city. For example, when the C.S. Mott Foundation paid to move the farmers' market closer to the city bus terminal to make it more accessible and more prominent, that resulted in opposition and a petition drive.[48]

Furthermore, the county health department also had not voiced concern about the levels of lead in the water or in the blood of children. My own college's public health team included people who had worked in county health departments, and so we thought that if there was a serious problem, the local health department would be a leading voice of concern. In all likelihood, though, the local health department in Genesee County was trapped in politics between the city, the county, and the state.

Neither the leadership of the College of Human Medicine nor its Public Health Division was at pediatrician-researcher Hanna-Attisha's first news conference on this issue, which shows we were slow to visibly

[47] K. Kozlowski, "UM declined to team with lead expert in Flint," *The Detroit News* (February 18, 2016). Available at: www.detroitnews.com/story/news/michigan/flint-water-crisis/2016/02/18/um-flint-lead-expert/80586022/. Accessed on: May 5, 2016.

[48] T. Moreno, "Not everyone excited about Farmer's Market move," *The Burton View* (March 21, 2013). Available at: http://burtonview.mihomepaper.com/news/2013-03-21/News/Not_everyone_excited_about_Farmers_Market_move.html. Accessed on: May 5, 2016.

acknowledge the crisis as significant. In our defense, we had not yet seen the data, and we had not yet learned how poor the science and coordination of this water source switch had been. In contrast to my own university and college, Hurley Medical Center, the local hospital where Hanna-Attisha practices, was there, and that institution has been stalwart in its support of her.[49]

Who Worked Effectively?

There were many local activists who tirelessly worked to bring attention to the Flint water problems. They include Lee Ann Walters, the mother of the harmed twins who pushed scientists to investigate. They also include a local pastors' group that attempted to bring attention to the overall water quality. Notably, in the end, scientists and pediatricians had to become activists in order for the water to be switched back to Detroit's system. The activist scientists and doctors then were able to draw in high-visibility journalists and commentators, including Rachel Maddow,[50] whose coverage of the topic relied largely on the work of intrepid, Michigan-based investigative journalists like Curt Guyette and made it impossible to ignore as a national shame.[51]

Of course, many national journalists and commentators have had a very simplistic view of the issues; they have tended to focus on two heroes, Mark Edwards – who is invariably named as a researcher and always identified with his Virginia Tech affiliation – and Mona Hanna-Attisha – who by contrast is generally identified as a pediatrician, usually without her hospital or academic affiliation. Left out of the coverage are the citizen-activists, lawyers, and local journalists who struggled for months to bring more attention to the issue. Also left out are the coauthors of the keystone *American Public Health Journal* article (Jenny LaChance, Richard Casey Sadler, and Allison Champney

[49] See R. Erb, "Flint doctor makes state see light."

[50] T. Dokoupil, "Maddow examines the catastrophic toll of the Flint water crisis," *MSNBC* (January 27, 2016). Available at: www.msnbc.com/msnbc/maddow-examines-the-catastrophic-toll-the-flint-water-crisis. Accessed on: October 20, 2016.

[51] "ACLU of Michigan reporter Curt Guyette named Michigan journalist of the year," American Civil Liberties Union of Michigan (January 29, 2016). Available at: http://aclumich.org/article/aclu-michigan-reporter-curt-guyette-named-michigan-journalist-year. Accessed on: May 5, 2016.

Schnepp) first-authored by Hanna-Attisha.[52] One of those coauthors, the health geographer Rick Sadler, was our new faculty member when the crisis broke. He was energetic, community-focused, and personally from the Flint area. He did the critically important geospatial analysis in the article that showed the relationship between areas of high water lead levels and children with high blood lead levels.[53] If the state had done that work, the Flint Water Crisis might have been over much sooner.

The email records released so far[54] do not document what Governor Snyder said to his department heads, nor what the department heads said to their staff, but we already know this: In the end no one wanted to deal with more bad news coming from Flint. As an administrator myself, I suspect that there was a systematic discomfort with bringing unwanted news forward. At senior administration levels, you don't even have to attack someone for people to become afraid of bringing you bad news. Even a stray remark, particularly one misheard, can accidentally change the character of your relationships and cause people not to tell you what you need to be told bluntly.

In many ways, this is a classic systems failure: People with important information did not share it or act on it. Clearly, the people of Flint, and all of us who have yet to experience something like Flint, would benefit from the government adopting the expectations of a "safety culture," in which people anywhere along a hierarchy can speak up without fearing the consequences of being wrong. The goal of a safety culture is not to fail to hold accountable those who are willfully responsible for mistakes, but rather to create a system where everyone is focused on safety rather than hierarchy, ego, credit, and blame.[55]

One of the problems with the government scientist system, as evidenced by the Flint Water Crisis, is that government scientists do not enjoy the freedoms and protections they need to sound alarms. By contrast, Edwards, Hanna-Attisha, Sadler, and the other involved university-affiliated researchers have enjoyed a system of academic

[52] M. Hanna-Attisha, J. LaChance, R.C. Sadler, and A. Champney Schnepp, "Elevated blood lead levels in children associated with the Flint drinking crisis: A spatial analysis of risk and public health response," *American Journal of Public Health*, 106:2 (February 2016), 283–290. Available at: http://ajph.aphapublications.org/doi/abs/10.2105/AJPH.2015.303003. Accessed on: October 21, 2016.

[53] M. Hanna-Attisha et al., "Elevated blood levels in children."

[54] See S. Roy, "MDEQ mistakes and deception."

[55] Institute of Medicine, *To Err Is Human: Building a Safer Health System* (Washington, DC: The National Academies, 1999).

freedom to protect them, at least to some degree, from the personal economic hardships of a political battle. That's simply not true for the people who work for the county, state, or federal governments. Because of the politicization of science in the last couple of decades, it can be extremely difficult for governmental scientists to talk about climate change,[56] gun violence,[57] sexual health and apparently the wholesomeness of drinking water without fear of retribution, including in the form of funding cuts. That means we depend on either a small number of very brave people like Del Toral or the continuation of academic freedom of people like the researchers who made the Flint Water Crisis impossible to ignore.

I believe the Flint Water Crisis ultimately serves to remind medical and public health workers how easy it is to make a consequential mistake when you don't put the health of others as your top priority. You don't have to be a bad person to make a terrible decision. One of the great mysteries of the Flint Water Crisis, the Tuskegee study,[58] and many of the ethical travesties documented throughout this book and elsewhere is why some people stand up for what is right while others do not. It may be the case that there are some people more easily moved to action, and some who are constitutionally unable to keep their mouths shut when they see a disaster in the making.

For me, as a medical educator and the interim dean of a medical school, this last question – "Why do some people stand up when we need them to?" – seems like the most important question of the day. If we can figure out why some people stand up, or even *who* will stand up, then perhaps we can train more professionals to do so.

56 A.C. Revkin, "Climate expert says NASA tried to silence him," *New York Times* (January 29, 2006). Available at: www.nytimes.com/2006/01/29/science/earth/climate-expert-says-nasa-tried-to-silence-him.html. Accessed on: October 20, 2016.

57 S. Ferris, "GOP blocks Dem attempts to allow federal gun research," *The Hill* (July 7, 2016). Available at: http://thehill.com/policy/healthcare/286847-gop-blocks-dem-attempts-to-allow-federal-gun-research. Accessed on: October 20, 2016. T.C. Frankel, "Why the CDC still isn't researching gun violence, despite the ban being lifted two years ago," *Washington Post* (January 14, 2015). Available at: www.washingtonpost.com/news/storyline/wp/2015/01/14/why-the-cdc-still-isnt-researching-gun-violence-despite-the-ban-being-lifted-two-years-ago/. Accessed on: October 20, 2016.

58 See S.M. Reverby, *Examining Tuskegee*.

INDEX

Books in the Series

Marcus Radetzki, Marian Radetzki and Niklas Juth
Genes and Insurance: Ethical, Legal and Economic Issues
Ruth Macklin
Double Standards in Medical Research in Developing Countries
Donna Dickenson
Property in the Body: Feminist Perspectives
Matti Häyry, Ruth Chadwick, Vilhjálmur Árnason and Gardar Árnason
The Ethics and Governance of Human Genetic Databases: European Perspectives
Ken Mason
The Troubled Pregnancy: Legal Wrongs and Rights in Reproduction
Daniel Sperling
Posthumous Interests: Legal and Ethical Perspectives
Keith Syrett
Law, Legitimacy and the Rationing of Health Care
Alastair Maclean
Autonomy, Informed Consent and the Law: A Relational Change
Heather Widdows and Caroline Mullen
The Governance of Genetic Information: Who Decides?
David Price
Human Tissue in Transplantation and Research
Matti Häyry
Rationality and the Genetic Challenge: Making People Better?
Mary Donnelly
Healthcare Decision-Making and the Law: Autonomy, Capacity and the Limits of Liberalism
Anne-Maree Farrell, David Price and Muireann Quigley
Organ Shortage: Ethics, Law and Pragmatism
Sara Fovargue
Xenotransplantation and Risk: Regulating a Developing Biotechnology
John Coggon
What Makes Health Public? A Critical Evaluation of Moral, Legal, and Political Claims in Public Health
Mark Taylor
Genetic Data and the Law: A Critical Perspective on Privacy Protection
Anne-Maree Farrell
The Politics of Blood: Ethics, Innovation and the Regulation of Risk
Stephen Smith
End-of-Life Decisions in Medical Care: Principles and Policies for Regulating the Dying Process

Michael Parker
Ethical Problems and Genetics Practice

William W. Lowrance
Privacy, Confidentiality, and Health Research

Kerry Lynn Macintosh
Human Cloning: Four Fallacies and Their Legal Consequence

Heather Widdows
The Connected Self: The Ethics and Governance of the Genetic Individual

Amel Alghrani, Rebecca Bennett and Suzanne Ost
Bioethics, Medicine and the Criminal Law Volume I: The Criminal Law and Bioethical Conflict: Walking the Tightrope

Danielle Griffiths and Andrew Sanders
Bioethics, Medicine and the Criminal Law Volume II: Medicine, Crime and Society

Margaret Brazier and Suzanne Ost
Bioethics, Medicine and the Criminal Law Volume III: Medicine and Bioethics in the Theatre of the Criminal Process

Sigrid Sterckx, Kasper Raus and Freddy Mortier
Continuous Sedation at the End of Life: Ethical, Clinical and Legal Perspectives

A. M. Viens, John Coggon and Anthony S. Kessel
Criminal Law, Philosophy and Public Health Practice

Ruth Chadwick, Mairi Levitt and Darren Shickle
The Right to Know and the Right Not to Know: Genetic Privacy and Responsibility

Eleanor D. Kinney
The Affordable Care Act and Medicare in Comparative Context

Katri Lõhmus
Caring Autonomy: European Human Rights Law and the Challenge of Individualism

Catherine Stanton and Hannah Quirk
Criminalising Contagion: Legal and Ethical Challenges of Disease Transmission and the Criminal Law

Sharona Hoffman
Electronic Health Records and Medical Big Data: Law and Policy

Barbara Prainsack and Alena Buyx
Solidarity in Biomedicine and Beyond

Camillia Kong
Mental Capacity in Relationship: Decision-Making, Dialogue, and Autonomy

Oliver Quick
Regulating Patient Safety: The End of Professional Dominance?

Thana C. de Campos
The Global Health Crisis: Ethical Responsibilities

Jonathan Ives, Michael Dunn and Alan Cribb
Empirical Bioethics: Theoretical and Practical Perspectives
Alan Merry and Warren Brookbanks
Merry and McCall Smith's Errors, Medicine and the Law (second edition)
Donna Dickenson
Property in the Body: Feminist Perspectives (second edition)
Rosie Harding
Duties to Care: Dementia, Relationality and Law
Ruud ter Meulen
Solidarity and Justice in Health and Social Care
David Albert Jones, Chris Gastmans and Calum MacKellar
Euthanasia and Assisted Suicide: Lessons from Belgium
Muireann Quigley
Self-Ownership, Property Rights, and the Human Body: A Legal and Philosophical Analysis
Françoise Baylis and Alice Dreger
Bioethics in Action
John Keown
Euthanasia, Ethics and Public Policy: An Argument against Legislation (second edition)
Amel Alghrani
Regulating Assisted Reproductive Technologies: New Horizons